Condition Monitoring in Thermal Power Stations

Condition monitoring (colloquially, CM) is the process of monitoring machinery to identify a significant change, which might be indicative of an underlying aberration. *Condition Monitoring in Thermal Power Stations: Case Studies* covers 25 case studies of all major areas of thermal power stations, which have suffered a lot of machinery condition issues. It covers the information about how machine vibrations can be brought down to the acceptable limits via the employment of condition monitoring techniques. It also includes detection methodology for condition monitoring deviations of all areas in a plant and suggested remedies.

Features:

- Covers pre-maintenance history to work execution through analysis and having the results with post-condition monitoring data.
- Uses actual machine maintenance photos for better understanding and visualization of case studies.
- Provides knowledge of each condition monitoring technique through all case studies.
- Helps readers to replicate the idea of case study diagnosis into plant/machinery of any type of industry.
- Illustrates real problems and real solutions for practical issues.

This book is aimed at professionals and researchers in plant engineering and maintenance, thermal power plants, and condition monitoring.

Condition Monitoring in Thermal Power Stations
Case Studies

Hemant Bari and Jignesh Thaker

CRC Press
Taylor & Francis Group
Boca Raton London New York

CRC Press is an imprint of the
Taylor & Francis Group, an **informa** business

First edition published 2025
by CRC Press
2385 NW Executive Center Drive, Suite 320, Boca Raton FL 33431

and by CRC Press
4 Park Square, Milton Park, Abingdon, Oxon, OX14 4RN

CRC Press is an imprint of Taylor & Francis Group, LLC

ISBN: 9781032531991 (hbk)
ISBN: 9781032558592 (pbk)
ISBN: 9781003432616 (ebk)

DOI: 10.1201/9781003432616

Typeset in Times
by codeMantra

Contents

Preface

This book has been produced to provide industrial cases for engineering education in the field of condition monitoring. This book mainly describes various condition monitoring techniques adopted at Adani Dahanu Thermal Power Station-ADTPS (capacity: 2 × 250 MW) and summarizes the problems diagnosed from April 2008 to May 2020. These utilized techniques have significantly helped to detect the failure of equipment at an early stage, which in turn avoided serious catastrophic failures. Condition monitoring describes a range of key techniques used in predictive maintenance engineering, and it has specific advantages such as reduction in unplanned outages, life extension of plant auxiliaries, better deployment of maintenance crew, advance planning in procurement of spares, reduction in secondary damages and increase in productivity.

Condition monitoring generally offers diagnostics, information and data for root cause analysis. It is used as a proactive defect elimination strategy to drive failure prevention. The main objectives of adopting a condition monitoring technique are to assess the condition of a machine, which can eliminate failures due to component degradation and identify the root cause of abnormalities which can extend the life of machine components that are absorbing the stresses and fatigue forces.

This book contains a basic understanding of various condition monitoring techniques and their effective utilization at ADTPS. In total, 25 real-life case studies of critical auxiliaries of ADTPS have been discussed extensively by considering all areas of power station, which have suffered from machinery condition problems related to high vibration, acceleration, noise, ultrasonic, distress level, oil parameter deviation, etc. These practices have saved the plant from huge losses in terms of generation, unplanned outages, breakdown time and associated maintenance costs. For remedy, the author used FFT spectrum and time waveform analysis, circle plot and phase analysis, Ferrography analysis and history data trending of the last 12 years. For better and quick understanding of reader, a large number of real photographs, spectrums, figures and tables are prepared and presented in this book. This will serve as a useful tool for maintenance engineers to become more proactive and accurate, for operation engineers to get more confidence regarding equipment reliability and for research scholars to know the importance and opportunities in the field of condition monitoring maintenance practices which is today's vital need of every industry. This book would also serve as a useful reference manual for practicing engineers/entry-level analysts.

In conclusion, a unique approach is presented in this book for planning and utilizing the various condition monitoring techniques at ADTPS in such a way that the machine/equipment always stays under observation by one or other condition monitoring techniques. The schedule of all condition monitoring tasks has been prepared in order to ensure timely monitoring of all major equipments at ADTPS.

Acknowledgments

The authors wish to acknowledge the management of Adani Dahanu Thermal Power Station (ADTPS) for providing the opportunity to compile and publish a case-study-based database of condition monitoring practices from the past 15 years. This book would not have been possible without the invaluable support of Mr. Rajendra Nandi, Chief Operating Officer (COO) at ADTPS, whose guidance and assistance were crucial to the completion of this project.

The authors also express their deep gratitude to the management of Adani University for providing the necessary infrastructure facilities needed for this project. Special thanks are extended to Prof. Ravi P. Singh, Provost of Adani University, for his continuous encouragement and support throughout this endeavor.

The authors would like to express their sincere thanks to Mr. Harish Pandey, Head of Operations and Maintenance, for his critical remarks and valuable suggestions. They are profoundly grateful to the maintenance and operations engineers for their unwavering support throughout the work and analysis.

Special thanks are also extended to Mr. Dinesh Mantri, Mr. Bhaven Sheth, Mr. Suhas Patil (Heads of Departments), and Mr. Atul Deshpande (Senior Analyst) for their guidance and continuous support. The authors also thank everyone who contributed, directly or indirectly, to the completion of this book, titled "Condition Monitoring in Thermal Power Stations: Case Studies"

Last but not least, the authors offer their praises and thanks to the Almighty God, the source of all knowledge and wisdom.

About the Authors

Hemant Bari is a General Manager in the Department of Maintenance Planning - Condition Monitoring Cell at Adani Dahanu Thermal Power Station (ADTPS). Presently, he is associated with data collection and analysis of all Condition Monitoring activities being executed during routine and overhaul period at ADTPS. He has completed the specialization course on Machine Diagnostic and Condition Monitoring from IIT Delhi after completing his BE in Production Engineering from the University of Mumbai. He has completed Vibration Analysis Level Three from Mobius Institute, Australia and certified with level 2 for Ultrasound Technology (Structure and Airborne) from the USA. He has completed PGDBM in Business Management from Emeritus University, Singapore. He is a level 2 certified NDT expert from ASNT for LPT, MPT, UT, PAUT and TOFD techniques. In addition, he is certified with Thermography and Oil Analysis level 1 certification.

Mr. Hemant is a board member of Condition Monitoring Society of India and Life member of Tribology and Acoustical Society of India. He has more than 16 years of experience in the field of various Condition Monitoring Techniques such as vibration analysis, motor current analysis and wear debris analysis. He has published more than 38 papers and presented at various National and International Condition Monitoring forums. In addition, he has conducted more than 65 one/two-day workshops on basics of vibration with hands on practice, oil analysis, thermography and basics of NDT techniques with power plant case studies at various Industries, IITs and reputed Engineering Institutes/Universities in India.

Dr. Jignesh Thaker is the Deputy Dean of Academics and an Associate Professor at the Faculty of Engineering Sciences and Technology (FEST), Adani University. He completed his Ph.D. and post-graduation in Mechanical Engineering from Sardar Vallabhbhai National Institute of Technology (SVNIT), Surat. His commitment to excellence earned him a Gold Medal for securing the top rank in his postgraduate studies at SVNIT. He is also the recipient of the prestigious Rashtriya Vidya Gaurav Puraskar, awarded in New Delhi for his outstanding contributions to the field of education. Dr. Jignesh has been honored with the Young Scientist Award by the Science and Engineering Research Board (SERB), Government of India.

With over 15 years of experience in Fluid-Thermal Engineering across industry, research, and academia, Dr. Jignesh has published more than 35 articles in reputed journals, books, and conferences at both international and national levels. As an endorsed PhD supervisor, he is currently guiding two doctoral students.

His ongoing collaboration with Adani Group industries involves various consultancy projects, industrial training programs, and student internships, highlighting his vital role in connecting academia with industry.

In addition to his academic responsibilities, Dr. Jignesh is the technical lead and mentor at the Innovation Hub (AIIH) at Adani University. Here, he spearheads initiatives aimed at enhancing students' potential, skills, and knowledge, with the objective of building a robust pipeline of future leaders, entrepreneurs, and employees.

1 Introduction

1.1 WHAT IS CONDITION MONITORING – DEFINITION AND FUNDAMENTALS

Condition monitoring stands as a vital discipline within the realm of industrial maintenance, rooted in the proactive assessment of machinery and equipment health. At its core, condition monitoring involves the systematic tracking and analysis of various parameters, such as vibrations, temperatures, and lubricant conditions, to discern early signs of potential faults or degradation. By vigilantly observing and interpreting the changes in operational parameters, practitioners can predict, prevent, and rectify issues before they escalate into critical failures.

The data gathered through continuous monitoring serves as a valuable repository of insights into the current and historical state of an asset. This comprehensive understanding of a machine's evolution allows for informed predictions about its future performance and potential degradation. As a result, maintenance activities can be strategically scheduled based on these predictive analyses. These practices are generally called as predictive maintenance – a method centered on anticipating failures and proactively addressing maintenance needs.

In essence, condition monitoring serves as the diagnostic heartbeat of industrial operations, allowing for a strategic approach to maintenance. It shifts the paradigm from reactive responses to a proactive, data-driven methodology, enhancing reliability, extending equipment lifespan, and minimizing unforeseen downtime.

1.2 EVOLUTION OF CONDITION MONITORING

The historical evolution of condition monitoring unfolds as a captivating narrative, shaped by innovation, challenges, and a relentless pursuit of operational excellence. Commencing in the mid-20th century, the earliest attempts at machinery health assessment were marked by rudimentary methods, including visual inspections and basic measurements. These humble beginnings set the stage for a journey that would witness transformative milestones and technological leaps.

The 1960s marked a pivotal era with the introduction of vibration analysis, a breakthrough that enabled practitioners to decipher the language of machine vibrations, unlocking unprecedented insights into potential faults. Subsequent decades witnessed the integration of oil analysis, infrared thermography, ultrasound testing and acoustic emissions into the condition monitoring toolkit, broadening its scope and efficacy.

In the 21st century, the convergence of cutting-edge technologies, such as the Internet of Things (IoT), artificial intelligence, and machine learning, ushered in a new paradigm for condition monitoring. Sensors embedded in machinery now provide real-time data streams, facilitating predictive analytics and catalyzing a shift toward a holistic and comprehensive approach to machinery health.

DOI: 10.1201/9781003432616-1

The evolution of condition monitoring, therefore, is not just a technological progression but a testament to human ingenuity and the perpetual quest for enhancing industrial reliability. Through this historical lens, a profound understanding of how condition monitoring has metamorphosed into an indispensable discipline, reshaping the landscape of industrial maintenance practices.

1.3 BENEFITS OF CONDITION MONITORING

This section highlights positive outcomes derived from the diligent implementation of condition monitoring techniques.

- **Decreased maintenance costs:** One of the foremost benefits is a substantial reduction in maintenance costs. By leveraging condition monitoring, potential issues are identified at their incipient stages, allowing for targeted interventions before they escalate into major problems. This strategic approach minimizes the need for extensive, costly repairs and replacements.
- **Reduced downtime:** Condition monitoring serves as a proactive sentinel against unplanned downtime. By continuously tracking machine health, abnormalities and potential failures are identified early, enabling timely maintenance interventions. This not only prevents sudden breakdowns but also contributes to a significant reduction in downtime, optimizing overall operational efficiency.
- **Extended asset life:** The ability to detect and address issues in their early stages translates into a prolonged lifespan for critical assets. Condition monitoring allows for precise interventions, mitigating wear and tear before it progresses to irreversible damage. Consequently, equipment and components can operate at peak efficiency for an extended duration, maximizing the return on investment.
- **Cost savings on prematurely changed resources:** Condition monitoring enables data-driven decision-making regarding resource replacement. For instance, if a monitoring system detects a consistent increase in noise from a component preceding failure, an alert can be triggered. This not only helps in avoiding premature replacements but also optimizes the utilization of resources, leading to substantial cost savings.
- **Real-time data in the digital age:** The evolution of condition monitoring into the digital realm has revolutionized the landscape. Internet-enabled and wireless-connected sensors, coupled with advanced software, provide continuous real-time measurements. This wealth of data is seamlessly relayed to integrated software for analysis and the creation of historical datasets.
- **Quicker issue resolution:** Real-time condition monitoring tools expedite the identification of root causes for issues. The ability to access wireless sensors remotely, using smartphones or tablets, ensures that employees are instantly connected to real-time data. This not only accelerates issue resolution but also empowers maintenance personnel with timely and actionable insights.

- **Shift from reactive to predictive maintenance:** Condition monitoring facilitates a paradigm shift from a reactive maintenance approach to a predictive one. With 24/7 measurements and a continuous overview of machine health, facilities can proactively plan and schedule maintenance activities, eliminating the need for reactive, emergency responses.

1.4 CONDITION MONITORING TECHNIQUES

Condition monitoring manifests itself through a rich tapestry of techniques, ranging from time-honored traditions to cutting-edge technologies. In this exploration, the diverse approaches that compose the condition monitoring toolkit, each offering a unique lens into the inner workings of machinery have been discussed.

1.4.1 Vibration Analysis

Vibration analysis is a pivotal process involving the measurement of vibration levels and frequencies in machinery. This information is then meticulously analyzed to assess the health of the machine. Vibration analysis is a versatile tool, capable of detecting a spectrum of issues such as imbalance, bearing failures, mechanical looseness, misalignment, resonance, natural frequencies, electrical motor faults, bent shafts, and even cavitation. Remarkably, it has been estimated that vibration warnings can offer up to three months of lead time before an actual failure occurs.

1.4.2 Oil Analysis

Oil analysis stands as a routine method for scrutinizing the health of machinery lubricants, assessing oil contamination, and monitoring machine wear. The process involves a comprehensive analysis, including moisture analysis, particle counting, elemental analysis, acid/base numbers, viscosity measurement, and Fourier transform infrared spectroscopy to determine various parameters. Spectrographic oil analysis tests can break down the chemical composition of oil, predicting potential issues based on the elements present.

1.4.3 Infrared Thermography

Infrared thermography employs thermal imagers to detect radiation and convert it into temperature, providing a visual representation of temperature distribution. Widely used for monitoring electrical and mechanical conditions, infrared thermography helps identify abnormal heat patterns within machinery. This proactive approach aims to ensure that machinery is running smoothly by detecting potential defects or inefficiencies.

1.4.4 Ultrasound

Ultrasound testing proves valuable for both low- and high-speed mechanical applications and high-pressure fluid situations. Digital ultrasonic meters measure high-frequency signals emitted from bearings in real time, offering insights into bearing defects, lubrication intervals, and more. Often utilized alongside vibration analysis, ultrasound testing employs the shock pulse method for efficient condition monitoring.

1.4.5 Acoustic Emissions

ISO 22096 outlines the monitoring of energy through acoustic emissions. This technique focuses on vibrations at higher frequencies than traditional vibration analysis, detecting signals caused by cracking or impact. Acoustic emissions offer a unique perspective, examining high-frequency signals independently of transducer position, machinery speed, or the configuration of rolling elements.

1.4.6 NDT Techniques during Annual Overhaul

Non-destructive testing (NDT) is the lifeblood of a well-run facility. The purpose of NDT is to inspect a component in a safe, reliable, and cost-effective manner without causing damage to the equipment or shutting down plant operations. This is in contrast to destructive testing where the part being tested is damaged or destroyed during the inspection process. These techniques can be applied to a variety of plant inspections, including metallic and non-metallic materials. NDT techniques are critical to testing mechanical & Electrical systems and their associated components.

1.5 TYPES OF CONDITION MONITORING

Expanding the exploration, the various types and techniques of machine condition monitoring, each tailored to collect specific data. These types encompass offline condition monitoring, online condition monitoring, and route-based condition monitoring, each serving distinct purposes in the maintenance landscape.

1.5.1 Offline Condition Monitoring

The equipments selected in the offline system are checked at regular intervals and the onset of malfunction is tried to be diagnosed at an early stage. Offline condition monitoring finds its niche in scenarios involving less critical or semi-critical assets. It relies on periodic scanning to observe the current condition of machinery. Commonly associated with techniques such as vibration analysis and oil analysis, offline monitoring becomes practical when occasional checks suffice. For instance, in offline oil analysis, organizations often use sampling kits for onsite viscosity and water level tests, obtaining basic results. Some organizations employ semi-automated, in-house testing equipment to assess wear metals, nitration, oxidation, and additive depletion. Offline condition monitoring is preferred for the monitoring of more economical and not-so-critical rotating equipment.

1.5.2 Online Condition Monitoring

Online condition monitoring systems mean 24/7 monitoring of monitored rotating equipment. Online systems are preferred especially for monitoring critical assets.

Online condition monitoring marks a significant leap into continuous and real-time measurement, facilitated by wirelessly connected sensors mounted on machinery. Integrated software interprets the data, providing instant warnings based on techniques such as vibration analysis, acoustic emissions, ultrasound, and infrared thermography. The flexibility of online monitoring lies in its customization, with a variety of sensors and monitoring systems available. Machine type, bearing type, machine speed, components, and elements are crucial considerations when selecting an online monitoring system.

Once sensors are appropriately mounted and wirelessly connected, they feed real-time data to a remote monitoring system. This system often integrates multiple sensor types, offering a comprehensive snapshot of the asset's current condition. Real-time alerts can be configured for remote devices or email notifications. The term "portable machine diagnostics" aligns with this type of online condition monitoring, where portable equipment reads data from mounted sensors.

1.5.3 Route-Based Condition Monitoring

Route-based condition monitoring introduces a technique where technicians intermittently record data using handheld devices like infrared imaging cameras. This method, often employed to create trend patterns, serves as a preliminary step to determine if advanced analysis is warranted. Technicians follow predefined routes, capturing data points at intervals to build a comprehensive understanding of machinery health.

Route-based monitoring is particularly beneficial for assets where continuous monitoring may not be feasible or necessary. It acts as a strategic approach, allowing for trend analysis and informed decisions regarding the need for more in-depth diagnostics. In the evolving landscape of condition monitoring, these diverse approaches cater to the varied needs of industries, offering flexibility and precision in maintaining machinery health.

1.6 INTRODUCTION ABOUT ADANI DAHANU THERMAL POWER STATION

Navigating the intricate landscape of condition monitoring techniques and their diverse applications, it becomes evident that the combination of precision and flexibility is paramount in maintaining the better health of any machinery. Pivoting the focus to a real-world exemplar of these principles, Adani Dahanu Thermal Power Station (ADTPS), stands as a testament to the successful integration of advanced condition monitoring practices. A series of insightful case studies unravel how ADTPS has harnessed these techniques to optimize efficiency, minimize downtime, and ensure the sustained performance of its critical assets.

Nestled amongst the commendable power generation plants in the country, ADTPS stands as a pinnacle of technological prowess and sustained excellence. Its commercial operations commenced in January 1996, and since then, it has garnered a plethora of awards, solidifying its reputation as an industry leader. This power plant is not merely a facility; it's a testament to innovation, performance, and enduring sustainability.

ADTPS has two units, each with a capacity of 250 MW, seamlessly integrating cutting-edge technology to optimize its operational efficiency. Positioned approximately 120km north of Mumbai on the route to Ahmedabad, its strategic location places at the forefront of the nation's power generation landscape.

Fuelling the plants is a thoughtfully blended mix of Indian washed coal and imported coal, maintaining a harmonious ratio of 80:20. The indigenous coal is sourced from the South Eastern Coalfields Limited (Korba), while imports arrive from diverse countries, including Indonesia and South Africa.

The accolades bestowed upon ADTPS are not solely for its power generation capabilities. The power station has implemented integrated management systems, obtaining certifications for quality management, environmental management, occupational health and safety assessment studies, social accountability management, information security management, and energy management. Moreover, the accreditation for the ADTPS coal testing laboratory underlines its commitment to maintaining the highest standards across all facets of operation.

Crucially, ADTPS distinguishes itself by strategically adopting advanced condition monitoring as a cornerstone of its maintenance management practices. This proactive approach involves prioritizing equipment for close monitoring based on a criticality index, ensuring heightened plant availability through a spectrum of condition monitoring techniques. Delving deeper into ADTPS, the case studies will unravel how this commitment to innovation and precision manifests in real-world case studies, showcasing the transformative impact of advanced condition monitoring on the operational resilience of this distinguished power generation facility.

2 Case Study 1
Coal Mill 2AB: Main Reducer Gearbox Bearing Deterioration – May 2020

EXECUTIVE SUMMARY

This is a success story of the implementation of oil condition monitoring (OCM) program at ADTPS, wherein imminent deterioration of main reducer gearbox, intermediate and output gear bearings of high tension (HT) auxiliary Coal Mill – 2AB was detected. The oil analysis report stated that high wear particle concentration (WPC) was observed with the presence of copper alloy particles which was suggested for visual inspection and for bearing deterioration. Such diagnosing the root cause of abnormity in advance and timely action avoided major gearbox repair. The present case study demonstrated that routine OCM identifies problems within machinery components before they become catastrophic failures, which can incur costly repairs and production to a halt even before deviation is reflected in vibration, acceleration, noise, ultrasonic, distress level and any other process parameters.

On 20 March 2020, the wear debris analysis report stated that WPC value deviated from 586 to 1219 against a limit value of 1000 WPC with gear wear particles of size ranging up to 24 μm and bearing wear copper alloy particles of size ranging up to 72 μm.

On 23 March 2020, oil parameter deviation was discussed in the meeting and decided to inspect main reducer gearbox bearings during the upcoming maintenance opportunity. On 24 March 2020, main reducer gearbox vibration, acceleration, distress level, noise and ultrasonic readings were taken. No deviations were noticed and decided to keep under close monitoring.

On 2 May 2020, scheduled mill changeover was carried out from 2AB to 2CD, i.e., Coal mill 2CD was taken into service by stopping the working of Coal mill 2AB. On 7 May 2020, the main reducer gearbox internal inspection was done in view of recommendations received from Ferrography report. It has been observed that output bearing clearances were found on higher side as 0.5 mm against limit of 0.3 mm. In addition, micro pitting on rollers and scoring marks on inner race were observed on intermediate bearing.

On 8 May 2020, minor looseness was noticed in the intermediate gear spacer during the inspection of main reducer gearbox. No abnormality was noticed for input and output gears. On 9 May 2020, intermediate gear and bearings were replaced during maintenance work. Both output shaft bearings were also replaced. Oil filtration was done and taken in service again.

On 11 May 2020, coal mill trial was taken after executing the maintenance work. Main reducer gearbox vibration, acceleration, distress level, noise and ultrasonic readings were found within desirable limit. After filtration, impurities level in oil were reduced from 0.24% to 0.10%.

DOI: 10.1201/9781003432616-2

2.1 PRE-MAINTENANCE HISTORY

On 13 March 2020, as per the scheduled activities on quarterly basis, oil samples were collected from Main Reducer Gearbox for wear debris analysis (as shown in Figure 2.1).

On 20 March 2020, the wear debris analysis report stated that wear particle concentration (WPC) value deviated from 586 to 1219 against the limit value of 1000 WPC with gear wear particles of size ranging up to 24 µm and bearing wear copper alloy particles of size ranging up to 72 µm (see Figure 2.2).

On 21 March 2020, based on the wear debris analysis report, work order was raised in SAP (System Application and Product in data-processing) software to mechanical maintenance department for internal inspection of main reducer gearbox. On 23 March 2020, oil parameter deviation was discussed in HoD meeting and decided to inspect main reducer gearbox bearings during opportunity (Figure 2.3).

Meanwhile, readings of main reducer gearbox (such as vibration, acceleration, distress level, noise and ultrasonic) were taken as shown in Figures 2.4–2.7. No major deviations were noticed from collected readings.

SKF make CMXA 75 vibration analyzer was used to collect vibration and acceleration data. Raython make RT 5001 noise meter used to collect noise data. Holroyd make MHC bearing checker was used to collect distress level and UE system make Ultraprobe 100 was used to collect ultrasound data.

After recording the database of vibration, acceleration, distress level, noise and ultrasonic, it was decided to keep the machine under close monitoring because

FIGURE 2.1 Oil sample collection.

PREDICT Technologies
EQUIPMENT SPECIFIC ANALYSIS
NAGARATNAM RESIDENCY,
PLOT NO-56, 5th FLOOR,
5th PHASE, K.P.H.B. COLONY
HYDERABAD 500085.
Telangana State

TEST REPORT NO:TR/PTI/03/20/833/20
CLIENT: ADANI ELECTRICITY MUMBAI LIMITED
ADANI DAHANU THERMAL POWER STATION-DAHANU

MARGINAL

SAMPLE ID: AEML-D/CM 2AB MR/2863/20/03/12408
SAMPLE DESC: COAL MILL 2 AB MAIN REDUCER
COMPONENT: GEAR BOX
LUBRICANT: NOT AVAILABLE

SAMPLE DATE: 13-03-2020
REC'D DATE: 17-03-2020
REPORT DATE: 20-03-2020
NO OF SAMPLES: 27
1st SAMPLE: 17-07-2008

RECOMMENDATION

Report is rated marginal due to the presence of marginal quantities of sand/dirt particles and high wear particle concentration. Change/Clean the lubricant (Filtration to remove the existing wear particles and contaminants) to avoid the generation of secondary wear. Check and arrest the possible inlets (damaged/broken seals etc) of sand/dirt ingress into the system.

Discussion of results

The wear particle concentration (WPC) is 1219.0. WPC limit is considered as 1000.0.
Normal rubbing wear particles are observed in marginal quantities.
Gear wear particles of size ranging up to 24 microns are observed in negligible quantities. Bearing wear particles of size ranging up to 28 microns are observed in small quantities. Medium alloy steel bearing wear particles of size ranging up to 24 microns are observed in small quantities. Copper alloy particles of size ranging up to 72 microns are observed in small quantities. Red oxides are observed in small quantities. Sand/dirt particles are observed in marginal quantities.

FIGURE 2.2 Wear debris analysis report.

adani | Electricity **Minutes of Meeting**

Adani Electricity Mumbai Limited	Minutes	Meeting

Number of ATR's	1	Number of DTR's	--	Number of ITR's	3

Meeting Meta Data

Meeting Level	Middle Management, Sr. Management	Function	DTPS	Department	MTP Maintenance and Planning
Meeting Type	Daily Planning Meeting	Meeting ID	DPM-23-March-2020	Category	Daily

No.	Cat ~ Sub Cat	Minute Narrative	Decision Taken	ID	Pr	Pri. Resp.	Assign To	Due Dt
1	Daily Defects Review ~ 1	Near miss: Dumper no 11 damaged the electrical pole near holding pond	CAPA to be taken by CHP-Operation	A-1	L	Paresh Deshpande	Paresh Deshpande	26/03/20
2	Daily Defects Review ~ 2	Coal Mill-2AB Wear debris particle concentration (WPC) deviated from 586 to 1219 against limit as 1000	Main reducer Gear box Oil to be replaced by Mechanical	I-1	--	--	--	--
3	Daily Defects Review ~ 3	Booster Fan# 01 Vibration maintaning on higher side from 3.8 to 4.44 mm/sec	kept under observation as presently remains steady.	I-2	--	--	--	--

FIGURE 2.3 Minutes of HoD meeting.

deviation was noticed only in oil parameter. On 2 May 2020, mill changeover was done from 2AB to 2CD proactively to check the root cause for deviation in oil parameter. On 7 May 2020, with reference to wear debris oil analysis report, internal inspection of main reducer gearbox was performed and completed.

FIGURE 2.4 Vibration and acceleration data collection at main reducer gearbox.

FIGURE 2.5 Noise data collection at main reducer gearbox.

FIGURE 2.6 Distress level measurement at main reducer gearbox.

FIGURE 2.7 Ultrasonic measurement at main reducer gearbox.

2.2 FERROGRAPHY/WEAR DEBRIS OIL ANALYSIS – BASICS

Ferrography is a specific type of oil analysis which mainly used to study the particle wear on machine components through analyzing the contaminants in lubricating oil. It can be used to predict and diagnose errors, occurring on machineries. It is closely related to tribology, which majorly focuses on friction between interacting surfaces. Wear debris analysis complements vibration analysis by providing independent and advanced indications of mechanical wear. Ferrography is a technique which is based on the systematic collection of oil samples from oil-lubricated machines. Ferrography analysis is microscopic evaluation of wear particles which identifies the type and severity of wear and generation mode that helps in understanding the source of the wear component (shown in Figure 2.8). When abnormal wear begins, no sharp increment in the concentration of small particles has been observed but there is a remarkable increment found in the concentration of large particles. Therefore, detection, measurement and analysis of these large wear particles can provide early and accurate information about the condition of the machine. It also helps in identifying the root cause leading to the wear out of components.

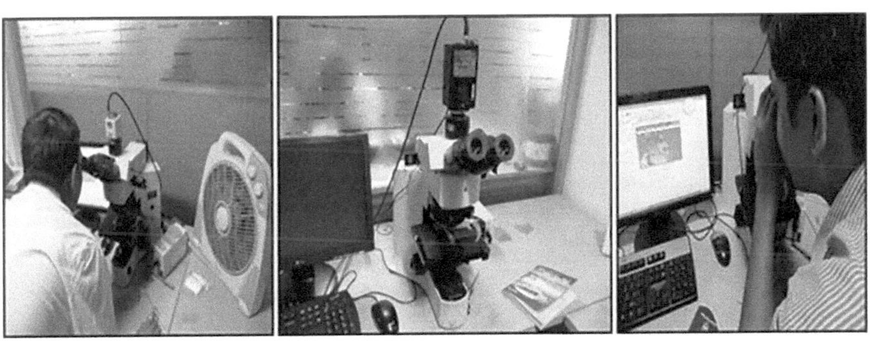

FIGURE 2.8 Ferrography analysis conducted at lab.

2.3 FERROGRAPHY/WEAR DEBRIS OIL REPORT – ANALYTICAL

Report is rated marginal due to the presence of marginal quantities of sand/dirt particles and high WPC. The WPC is found at 1219. WPC limit is considered as 1000. Normal rubbing wear particles are observed in marginal quantities. Gear wear particles of size ranging up to 24 µm are observed in negligible quantities. Bearing wear particles of size ranging up to 28 µm are observed in small quantities. Medium alloy steel bearing wear particles of size ranging up to 24 µm are observed in small quantities. Copper alloy particles of size ranging up to 72 µm are observed in small quantities. Red oxides are observed in small quantities. Sand/dirt particles are observed in marginal quantities (Figures 2.9–2.11).

Ferrography oil analysis recommends to clean/filter the lubricant for removing the existing wear particles and contaminants which helps to avoid the generation of secondary wear and inspection of gearbox internals.

2.4 OBSERVATIONS RECORDED DURING MAINTENANCE WORK

Condition of motor and coupling of main reducer gearbox were checked as shown in Figure 2.12. No deviation in readings was noticed. Main Motor drive end (DE)

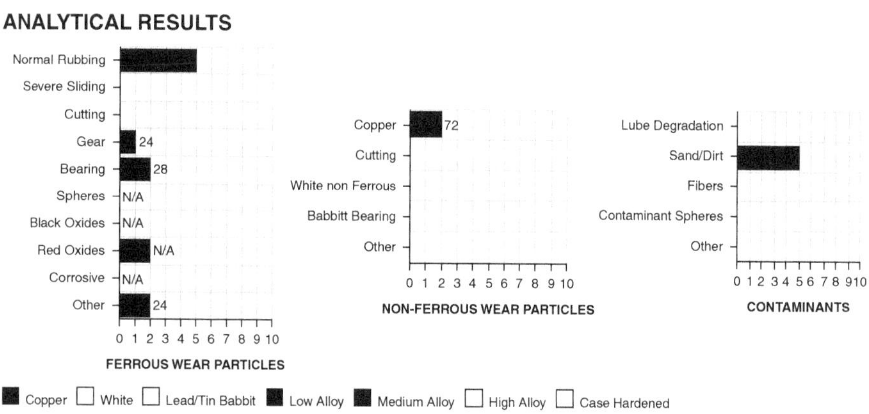

FIGURE 2.9 Wear particle findings.

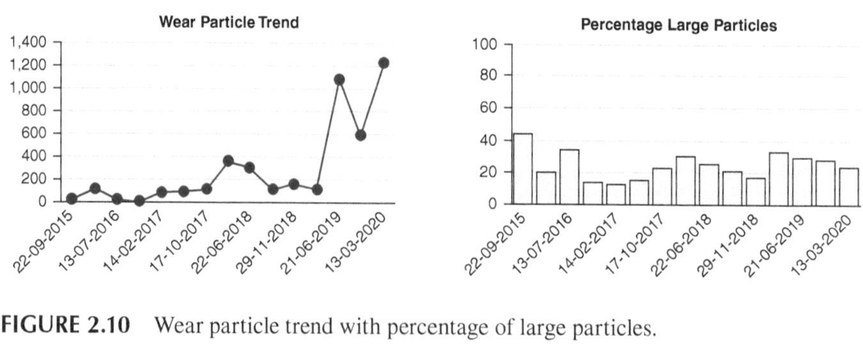

FIGURE 2.10 Wear particle trend with percentage of large particles.

FIGURE 2.11 Microscopic image of oil particles.

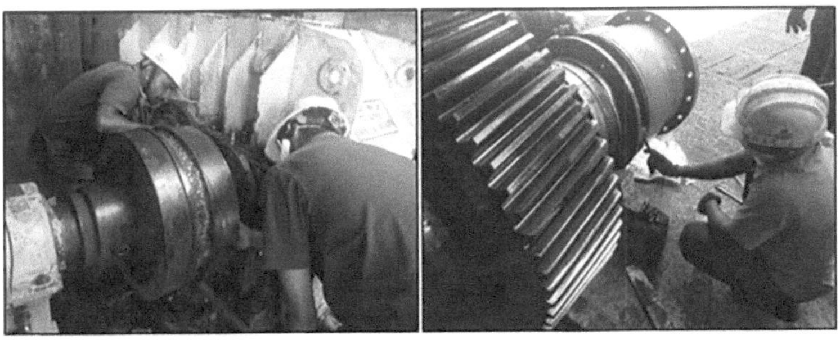

FIGURE 2.12 Motor coupling (LHS) and gearbox coupling (RHS).

FIGURE 2.13 Top (LHS) and bottom (RHS) half of motor DE bearing].

bearing inspection was done. All clearances were found within the limit as shown in observed from Figure 2.13. Main reducer gearbox internal inspection was done (as shown in Figure 2.14), and input bearing clearances were found within limit. No abnormality was noticed for gears. Output bearing clearances were found on higher side

FIGURE 2.14 Internal inspection of gears and bearings.

FIGURE 2.15 Output bearing clearance check.

as 0.5 mm against limit of 0.3 mm (see Figure 2.15). Micro pitting on rollers and scoring marks on inner race were observed on intermediate bearing (as shown in Figure 2.16). During a thorough inspection of main reducer gearbox, minor looseness was noticed in intermediate gear spacer (shown in Figure 2.17).

2.5 MAINTENANCE WORK PERFORMED

Main reducer gearbox is de-coupled from the rest of the drives (as shown in Figure 2.18). Main reducer gearbox top cover was removed as shown in Figure 2.19.

Oil drained and filtration was done for reducer gearbox as per oil sample report recommendation. Meanwhile, both oil samples (used and filtered) were sent for chemical analysis (Figure 2.20).

FIGURE 2.16 Intermediate gear and bearing clearance check.

Intermediate shaft was taken out for bearing removal and gear inspection (Figure 2.21).

By observing micro pitting on bearing and looseness in gear spacer, it was decided to replace the intermediate gear with spare serviced one and new bearings (Figure 2.22). Output gear shaft was taken out as shown in Figure 2.23. Output shaft coupling and bearings were removed as shown in Figure 2.24. Intermediate and output shaft cleaning was done, and shaft clearances were checked before the installation of new bearings (Figure 2.25). New bearings were installed on output gear shaft as shown in Figure 2.26. Similar coupling was installed on output gear shaft (Figure 2.27). Spare serviced intermediate gear and output gear with new bearings were kept back in gearbox (Figure 2.27). Main reducer gearbox top cover is placed as shown in Figure 2.28. Filtered oil filling of 420 liters was done in main reducer gearbox as shown in Figure 2.29. Main reducer gearbox was aligned with rest of drives (see Figure 2.30).

FIGURE 2.17 Internal inspection of main reducer gearbox.

FIGURE 2.18 De-coupling of main reducer gearbox.

FIGURE 2.19 Removal of top cover of gearbox.

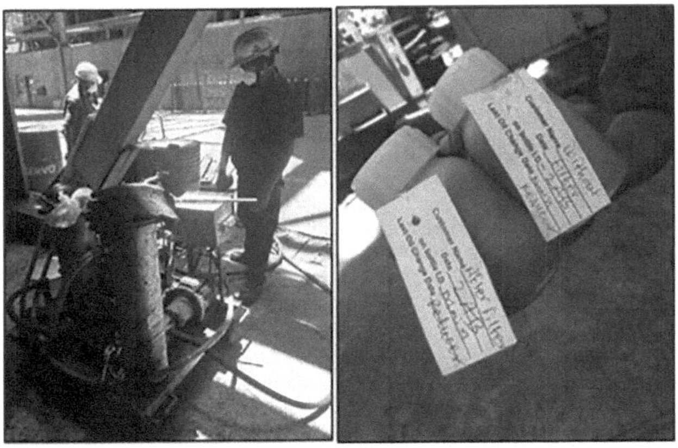

FIGURE 2.20 Gearbox oil filtration in progress.

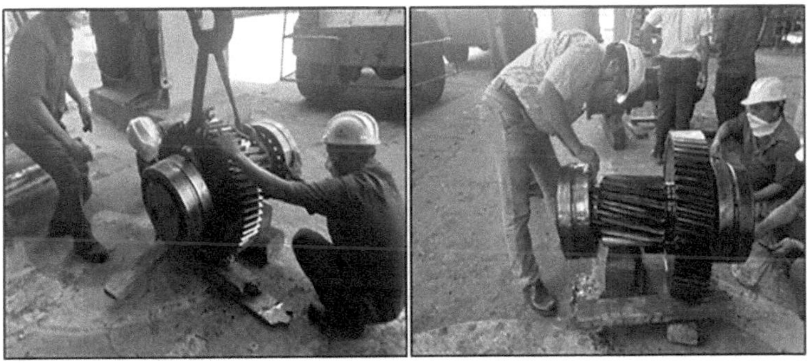

FIGURE 2.21 Intermediate gear and bearing replacement work.

FIGURE 2.22 Bearing installation for intermediate gear.

FIGURE 2.23 Removal of output gear.

FIGURE 2.24 Removal of coupling and bearings from output gear.

FIGURE 2.25 Pre-cleaning of output shaft bearing area.

FIGURE 2.26 New bearing installation for output shaft.

FIGURE 2.27 Installation of output gear in gearbox.

FIGURE 2.28 Gearbox top cover placement.

FIGURE 2.29 Oil filling process in gearbox.

FIGURE 2.30 Final alignment of gearbox with rest of drive.

On 11 May 2020, after maintenance work execution, Coal mill trial was taken. Vibration, acceleration, distress level, noise and ultrasonic readings of main reducer gearbox were found within limit. After the filtration process, impurities level in oil was reduced from 0.24% to 0.10%.

2.6 POST-MAINTENANCE OIL FILTRATION READINGS

On 9 May 2020, a chemical analysis report stated that reduction in impurities level after filtration was recorded from 0.24% to 0.10%. Means oil results were improved after filtration (Figure 2.31).

2.7 POST-MAINTENANCE CONDITION MONITORING PARAMETERS

Table 2.1 shows pre- and post-maintenance comparison for all condition monitoring parameters such as vibration, acceleration, distress level, noise and ultrasound level. All readings were found within limits.

adani
Electricity

Record Format No: 13.3b.14
Page: 9 of 10

ADANI DAHANU THERMAL POWER STATION
CHEMICAL DEPARTMENT

(COAL MILL 2AB - MAIN REDUCER GEARBOX)

(SERVO MESH - EE 460)

Sampling Date: 09/05/2020 Reporting Date: 09/05/2020

SR. NO.	PARAMETERS	Before Filtration	After Filtration
1.	Appearance	Thick, Dark Brown	Thick, Dark Brown
2.	Density gm/cc at 15°C	0.9140	0.9110
3.	Moisture in PPM	250.6	172.4
4.	Viscosity in cSt. at 40 °C	450	445
5.	Impurities %	0.243	0.107

LIMITS

PARAMETERS	SERVO MESH - EE 460
Moisture in ppm	1000
Viscosity in cSt at 40°C	325.0 to 625.0
Impurities (Max.) %	0.5

PREPARED BY CHECKED BY APPROVED BY

C.C.TO:- 1) USER DEPT.
2) MTP DEPT.

FIGURE 2.31 Pre-post oil filtration results.

TABLE 2.1
Pre-post Maintenance Condition Monitoring Parameters

Condition Monitoring Report of Coal Mill 2AB

			Initial Readings 02/05/2020		After Maintenance Work 11/05/2020	
Point		Position	Velocity (mm/s) RMS	Acceleration (g) Peak	Velocity (mm/s) RMS	Acceleration (g) Peak
Main motor	(NDE)	H	2.8	0.2	2.6	0.2
		V	1.4	-	1.9	-
		A	4.8	-	4.9	-
Main motor	(DE)	H	5.2	0.5	2.5	0.4
		V	5.6	-	3.4	-
		A	4.7	-	5.4	-
Reducer GB input bearing	(DE)	H	3.7	1.3	3.1	0.5
		V	2.5	-	2.1	-
		A	3.6	-	1.6	-
Reducer GB input bearing	(NDE)	H	4.8	1.7	3.7	1.5
		V	2.8	-	3.8	-
		A	3.5	-	2.4	-
Reducer GB intr. bearing	(NDE)	H	2.8	2.0	3.5	1.3
		V	1.3	-	3.3	-
		A	3.4	-	1.8	-
Reducer GB output bearing	(NDE)	H	3.4	1.6	2.4	0.9
		V	2.5	-	5.5	-
		A	4.0	-	4.3	-
Pinion bearing	(DE)	H	4.2	0.6	3.6	0.4
		V	5.4	-	5.8	-
		A	3.2	-	2.0	-
Pinion bearing	(NDE)	H	3.4	0.4	4.9	0.7
		V	4.6	-	5.1	-
		A	2.9	-	4.8	-
Distress level			**04 dst**		**04 dst**	
Ultrasonic level			**05 LEDs**		**04 LEDs**	
Noise level			**96.6 dB**		**91.7 dB**	

2.8 CONCLUSIONS

Coal mill is an important and critical auxiliary of thermal power station which used to pulverise coal in required size for efficient combustion in boiler. However, Coal mill supplies coal as a fuel to boiler, the reliability needs to be very high. Any failure associated with coal mill, directly affects to the generation loss of 0.125 Mus (million units) per hour which costs revenue of around Rs. 437,500 per hour. There are three nos. of coal mills at ADTPS of 50% capacity (i.e. two running at full load and other is in stand-by condition). As mentioned, coal mill is critical equipment in the system and if it fails during the non-availability of stand-by mill, it may lead to tripping of plant and interruption in power supply to our valued customers of Mumbai, India.

This case study has clearly demonstrated that oil condition monitoring (OCM) is a vital tool in rotating machinery analysis and for the detection of abnormality at an early stage even before deviation is reflected in vibration, acceleration, noise, ultrasonic, distress level and any other process parameters. High WPC with the presence of copper alloy particles of main reducer gearbox oil analysis was initiated a visual inspection and suspected that deterioration of bearings is in progress.

This helped in diagnosing the root cause of the abnormality well in advance and helped to avoid secondary damage to main reducer gearbox. This practice has saved the plant from huge losses in terms of unplanned breakdown time and associated maintenance, manpower and spare costs. Other major benefits include prevention of downtime, avoidance of major gearbox repair or replacement costs and reliability of unit.

3 Case Study 2 *Booster Fan 01 (FGD): Fan Bearings Looseness – April 2020*

EXECUTIVE SUMMARY

The present case study represents the success story of the implementation of condition monitoring techniques at ADTPS, wherein imminent Fan Bearing Looseness of HT auxiliary Booster Fan – 01 of Flue Gas Desulphurization (FGD) unit was diagnosed.

Vibration monitoring with Fast Fourier Transform (FFT) and time waveform spectrum analysis helped in diagnosing the exact root cause of abnormity well in advance. This has saved the plant from huge losses in terms of power generation loss, unplanned breakdown time, associated maintenance and spare costs and environment compliance with FGD availability as per statutory norms.

On 30 March 2020, both the FGD units got bypassed because of the sudden trip of seawater pumps. This tripping happened due to low protection of bearing cooling header pressure was in service, which can avoid any secondary damage to pump. After the restoration of sea water pumps, unit-02 FGD was taken in service, but unit 01 Booster Fan tripped twice due to high fan vibration when loading increased up to 50%.

On 31 March 2020, fan internal inspection was carried out. Fan was taken in service and gradually loaded. At 54% fan loading condition, fan casing online and offline horizontal vibrations were maintained on higher side up to 5.4 and 4.2 mm/s (RMS), respectively. Also, horizontal vibrations of Non Drive End (NDE) and Drive End (DE) of motor were recorded on higher side of 1.7 and 2.2 mm/s, respectively. No other deviations were noticed in any process parameters. With reference to the recorded spectrum analysis, it was decided to check the alignment between motor and fan.

On 1 April 2020, alignment was found disturbed by 23 divisions axially and same corrected by adjusting the motor position. On 2 April 2020, fan was again taken in service, because at 55% fan load condition, the vibration level was observed on higher side. On 3 April 2020, the motor position was restored to original, and fan was taken in service. But no improvement was noticed. Spectrum analysis revealed the predominant frequency 1×, 2× and 3× indicating misalignment during loading condition suspected due to looseness of fan bearing. This might have happened during sudden tripping and load throw-off condition for fan (access trust on fan due to impact of tripping). High-resolution time waveform and circle plot analysis confirmed the impacts at the running speed of fan with high crest factor of 4.7. Thus, it was confirmed that fan bearings were loosed, and deterioration was in progress. This problem was discussed in departmental meeting and decided to go for the replacement of fan bearings.

On 9 April 2020, the motor was aligned with fan, and the trial was taken at full load condition after replacing fan bearings. Fan casing online and offline horizontal vibrations were reduced from 5.4 and 4.2 to 2.8 and 1.7 mm/s respectively. Horizontal vibrations of

DOI: 10.1201/9781003432616-3

NDE and DE motor reduced from 1.7 and 2.2 to 0.4 and 0.5 mm/s respectively. During rotor maintenance, fan bearing float was found on higher side of 3 mm (limit value ranges between 0.05 and 0.10 mm). Through detailed inspection of fan rotor including all 16 nos. of blades, bearing housing, and shaft, no abnormality/secondary damage was noticed.

3.1 PRE-MAINTENANCE HISTORY

On 30 March 2020, all running sea water pumps tripped due to water leakage, which activated the protection of low header pressure of the bearing cooling system. This results in tripping of both the units of Flue Gas De-sulfurization (FGD) plants. After the restoration of sea water pumps, both units of FGD was taken in service but booster fan of unit 01 was tripped twice due to high vibration of fan while loading up to 50% capacity. This fan was kept off to inspect fan internals.

On 31 March 2020, fan internal inspection was done. Fan was started and gradually loaded by observing vibration level. At 54% loading, fan casing online and offline horizontal vibration readings were maintained on higher side of 5.4 and 4.2 mm/s, respectively (Figure 3.1).

Offline motor non drive end (NDE) and drive end (DE) horizontal vibration readings were observed on higher side of 1.7 and 2.2 mm/s, respectively (Figure 3.2).

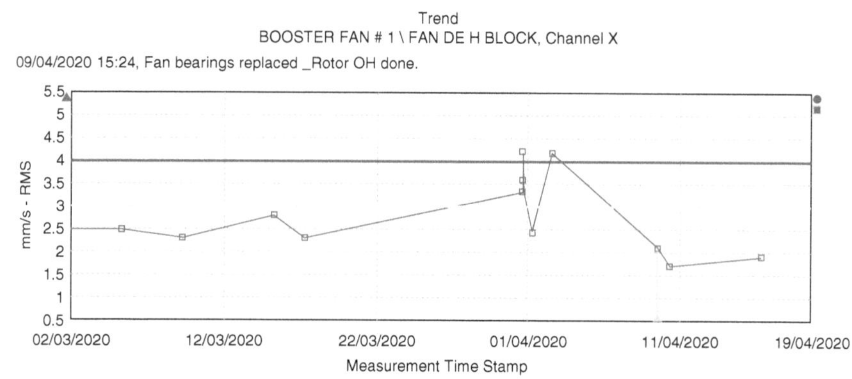

FIGURE 3.1 Vibration trend of fan casing block – horizontal.

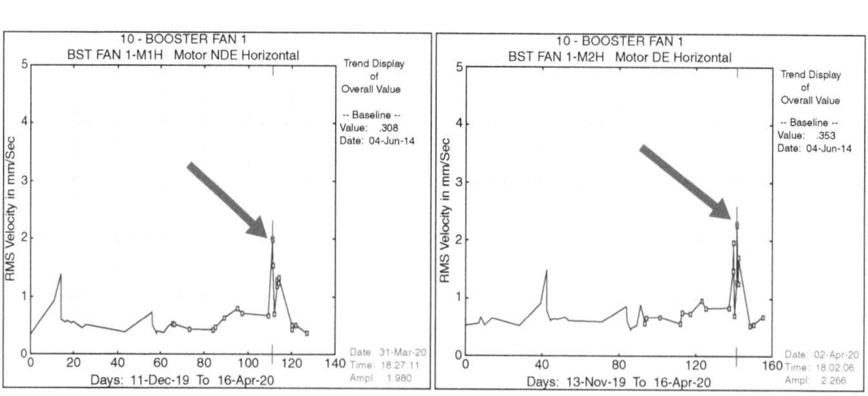

FIGURE 3.2 Vibration trend of motor NDE and DE bearings – horizontal.

No deviations were noticed in any process parameters. As per spectrum analysis, CMC (Condition Monitoring Cell) recommended to check alignment and SAP work order (300000127188) raised to mechanical maintenance department. On 1 April 2020, alignment was found to be disturbed by 23 divisions axially and same corrected by adjusting the motor position (represented in Figure 3.3).

On 2 April 2020, fan was again taken in service and observed that at 55% load condition, the vibration level was maintained on higher side. Readings were taken on the motor and fan foundation to check for any looseness (as shown in Figure 3.4). No deviation was noticed for the above readings which eliminated the possibility of foundation looseness. On 3 April 2020, the motor position was restored to original, and the fan was taken in service. But no further improvement was noticed (Figure 3.5). This high vibration issue was discussed in HOD meeting (Figure 3.6). By observing high fan bearing casing vibration trend, it was decided to go for bearing replacement.

FIGURE 3.3 Alignment work in progress.

FIGURE 3.4 Vibration data collection at motor and fan foundation.

FIGURE 3.5 Online vibration reading at fan casing block – horizontal.

adani | Electricity **Minutes of Meeting**

Adani Electricity Mumbai Limited	**Minutes**	Meeting	

Number of ATR's	1	**Number of DTR's**	--	**Number of ITR's**	10

Meeting Meta Data

Meeting Level	Middle Management, Sr. Management	**Function**	DTPS	**Department**	MTP Maintenance and Planning
Meeting Type	Daily Planning Meeting	**Meeting ID**	DPM-04-April-2020	**Category**	Daily

No.	Cat ~ Sub Cat	Minute Narrative	Decision Taken	ID	Pr	Pri. Resp.	Assign To	Due Dt
1	Daily Other Review ~ 3	BOQ for CCI Value and suction piping modification of vacuum pump for improvement plan execution.	Estimate to be done by Mechanical for 08.04.2020	A-1	M	Arun Patil	Arun Patil	08/04/20
2	Daily ADTPS Safety Review ~ 1	Sharing of fatality incidence in other power station - Metallic rod touched the overhead 11 KV line while carrying by worker during road construction.	Learnings discussed and necessary safety precautions to be ensured by concerned users.	I-1	--	--	--	--
3	Daily Defects Review ~ 1	Booster fan #01 high vibration issue	Work in progress by Mechanical.	I-2	--	--	--	--

FIGURE 3.6 Minutes of HOD meeting.

3.2 TIME WAVEFORM - BASICS

Figure 3.7 shows the featured time waveform plot which represents the signal that appears from sensor or probe when graphed as amplitude versus time. The time waveform shown in Figure 3.8 is a record of observing the path of amplitude (in milliseconds) when the shaft turns and gears mesh. Each minute (small to smallest) change that results from impacts, rubs, scrapes, rattles, surges, etc. is recorded in the time waveform and then summarized in the spectrum. Therefore, it is critical to record the time waveform correctly and analyze it when a fault condition exists. Time waveform analysis is especially useful while analyzing the gearboxes and bearings.

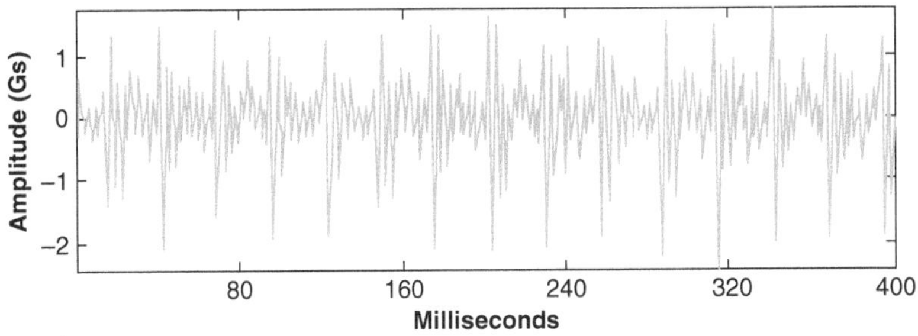

FIGURE 3.7 Basics of time waveform.

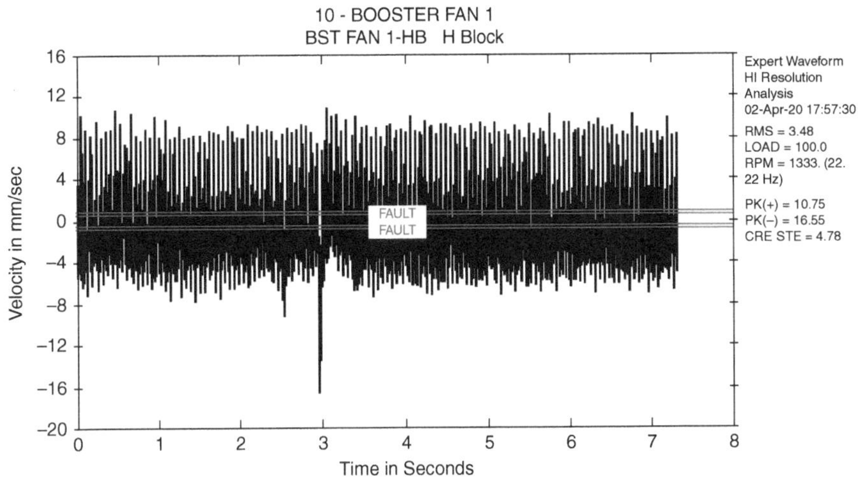

FIGURE 3.8 Fan casing time waveform spectrum.

3.3 TIME WAVEFORM - ANALYTICAL

Figure 3.8 is the time waveform spectrum, collected for fan bearing casing in the horizontal direction for finding the root cause of abnormality.

Zoomed waveform (shown in Figure 3.9) shows the time between any two impacts.

From Figure 3.9 (zoomed spectrum), selected any 02 impact times to calculate the defect frequency. Time difference between the two impacts was calculated as,

T = 2149.1 − 2089.1 = 60 ms = 0.06 seconds

Frequency in Hz, **F** = 1/T= 1/0.06 = 16.67 Hz

Frequency in CPM, **F** = 16.67 × 60 = 1000.2 cpm (co-relates with fan speed)

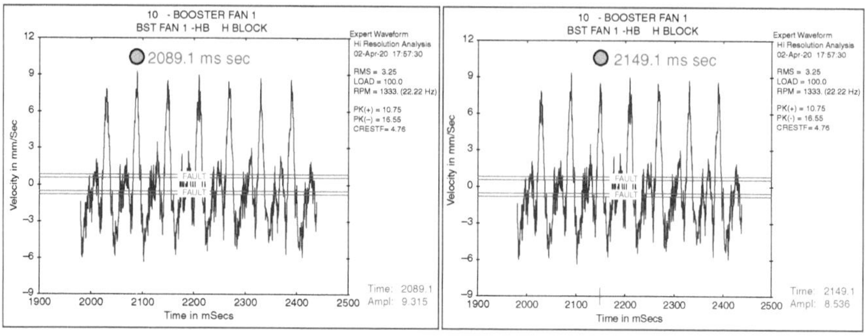

FIGURE 3.9 Zoomed time waveform spectrum.

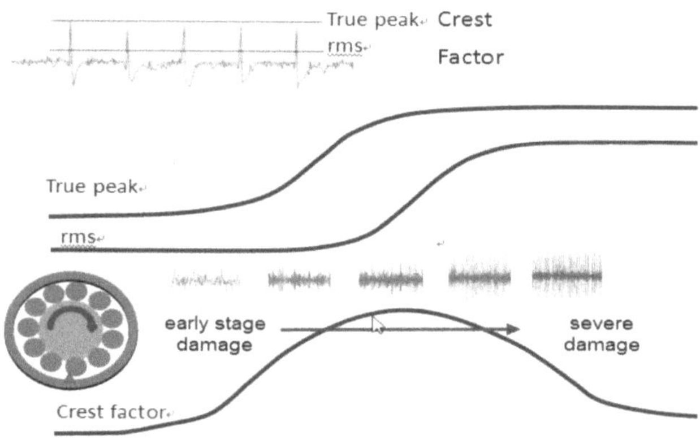

FIGURE 3.10 Basics of crest factor.

3.4 CIRCLE PLOT INTERPRETATION WITH CREST FACTOR

The circle plot wraps the waveform around in a circle. It provides the instantaneous database of vibration when the shaft rotates. This plot is critical for synchronous faults such as gear teeth, rubs, looseness and even sometimes misalignment, too. The crest factor of a waveform is the ratio of the peak value of the waveform to the RMS value of that waveform. It is also sometimes called the "peak-to-RMS-ratio". The crest factor of a sine wave is 1.414 (i.e. the peak value is 1.414 times the RMS value) (Figure 3.10).

A typical vibration signal from a machine with a large unbalance will have a crest factor near about 1.5, but as the bearings begin to wear, and impact starts to happen, the crest factor will become much larger than this value. The crest factor is one of the important measures for analyzing the machine condition. Figure 3.11 shows circle plot of fan bearing casing (Horizontal) with higher value of crest factor 4.7.

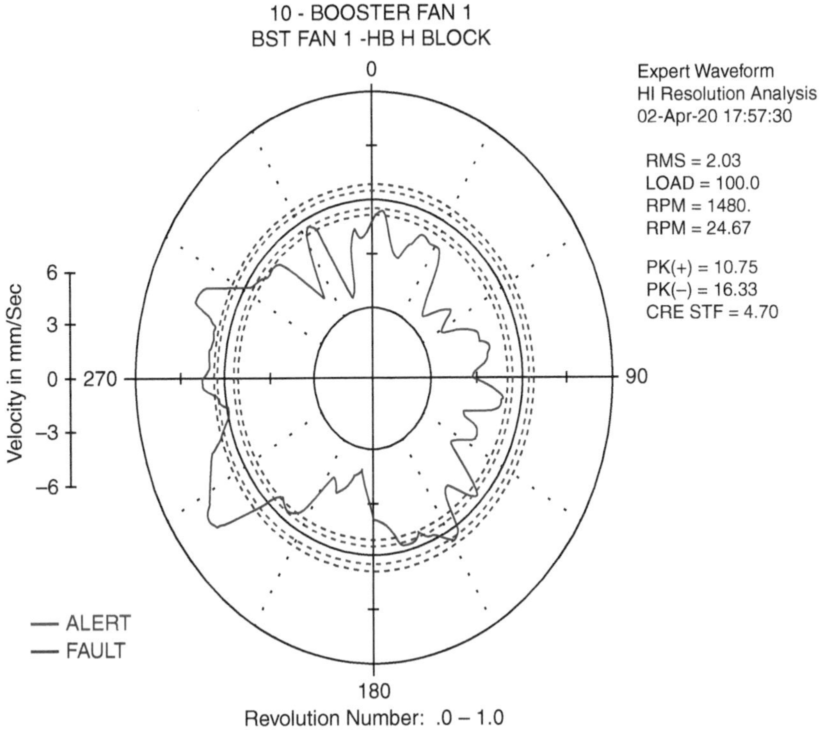

FIGURE 3.11 Circle plot of fan bearing casing (horizontal).

3.5 VIBRATION FFT SPECTRUM DIAGNOSIS

From vibration spectrum analysis, it was noticed that increase in vibration was observed only in the horizontal direction for motor and fan, both (refer Figures 3.12 and 3.13). FFT Spectrum analysis was carried out and found predominant vibration frequency as 1×, 2×, 3× … 8×. High-resolution time waveform spectrum showed the time between two impacts is 1000.2 rpm which co-relates with fan speed confirming bearing fault.

Circle plot analysis confirmed the impacts at the running speed of fan with high crest factor 4.7 (as shown in Figure 3.11). Thus, it was confirmed that fan bearings are loose, and deterioration is in progress.

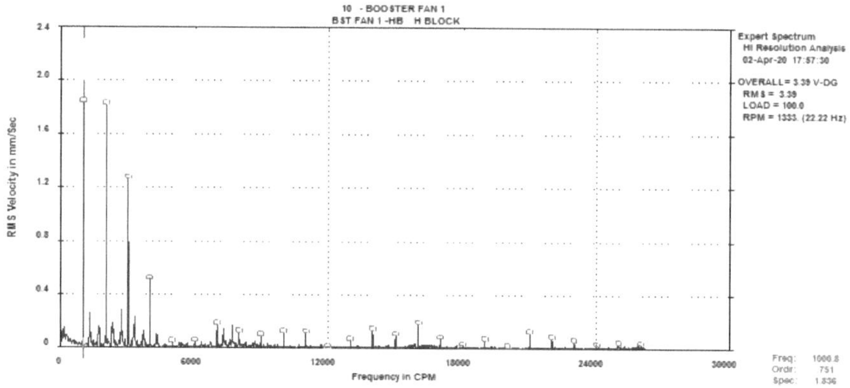

FIGURE 3.12 Vibration spectrum of fan casing (horizontal).

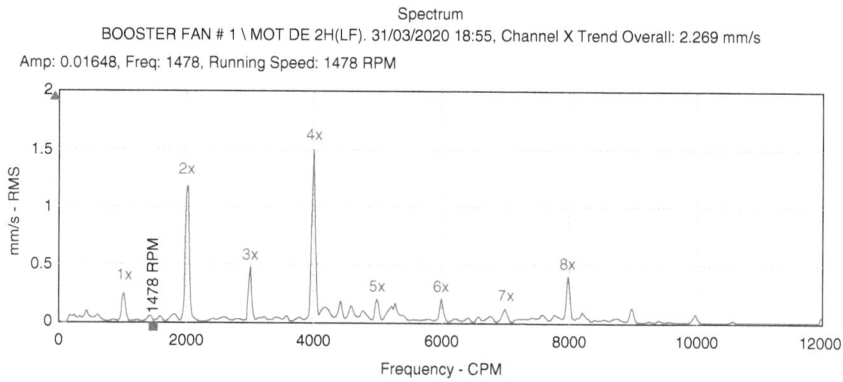

FIGURE 3.13 Vibration spectrum of fan casing (horizontal).

3.6 FREQUENCY ANALYSIS

While recording the online and offline vibration readings of fan casing, it was found that the readings were observed on higher side of 5.4 (online) and 4.2 mm/s (offline) during the loading process after sudden tripping of fan. Also, deviation in vibration was noticed in offline motor NDE and DE from 0.5 and 0.6 to 1.7 and 2.2 mm/s (RMS) respectively. FFT spectrum analysis revealed predominant vibration frequency as 1×, 2×, 3× … 8× which suspected bearing looseness. Time waveform and circle plot analysis interpreted impacts at 1000.2 rpm which co-relates with fan speed. High crest factor of 4.7 was confirmed fan bearing looseness and deterioration was in progress. It was concluded that vibration of fan bearing casing was increased due to internal impacts generated from loose bearing elements which might have happened during sudden tripping and load throw off condition for fan on 30 March 2020. Finally, it was decided to go for replacement of fan bearings.

3.7 OBSERVATIONS DURING MAINTENANCE WORK

Coupling condition of motor and fan side was checked as shown in Figure 3.14. No deviation was noticed. Fan bearing float was found on higher side as 3.0 mm against its limit value range of 0.05–0.10 mm (see Figure 3.15). All fan blades and fan internal inspection was carried out and no deviation was observed (as shown in Figure 3.16). Detailed inspection of fan rotor and bearing housing was carried out

FIGURE 3.14 Motor and fan coupling check.

FIGURE 3.15 Fan bearing float check.

FIGURE 3.16 Fan blade inspection by maintenance team.

FIGURE 3.17 Fan rotor inspection.

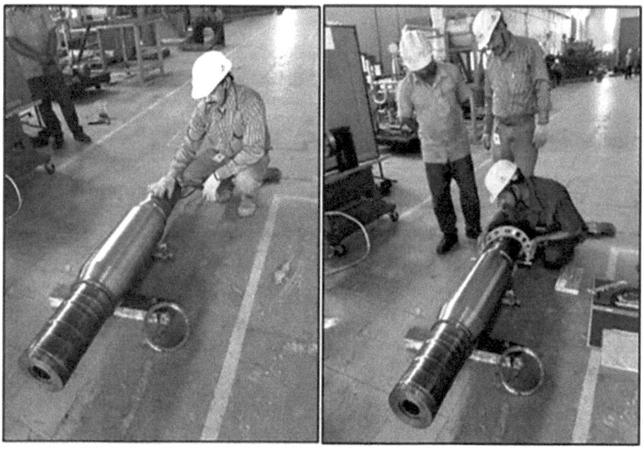

FIGURE 3.18 Fan shaft clearance check.

and no further deviations were observed (see Figure 3.17). Detailed inspection of fan shaft was carried out and no abnormality was noticed (see Figure 3.18).

3.8 PERFORMED MAINTENANCE WORK

First, motor was de-coupled from rest of the drive group. Before dismantling, initial alignment readings were taken for reference. All fan blades (16 nos.) were removed. Inspection and cleaning were carried out (see Figure 3.19). Fan rotor was taken out for bearing replacement as shown in Figure 3.20. Rotor dismantling was done, and all bearings (03 nos.) were removed (see Figure 3.21). Fan rotor servicing, bearing housing and shaft cleaning were carried out. During the maintenance, it was noticed that axial float of fan rotor shaft is on higher side of 3.0 mm due to increase in gap between spacers which was observed in between the angular contact bearing. This was the main root cause for sudden tripping of fan. Hence, care has been taken during fixing of impeller assembly on the shaft and verified by checking the float of the

FIGURE 3.19 Removal of all fan blades.

FIGURE 3.20 Fan rotor at maintenance bay.

FIGURE 3.21 Fan rotor servicing in progress.

shaft bearing assembly. New bearings (03 nos.) were installed on fan shaft and measured float was 0.05 mm (Figure 3.22). Rotor assembly was completed by installing the new bearings and placed at their location successfully (shown in Figure 3.23).

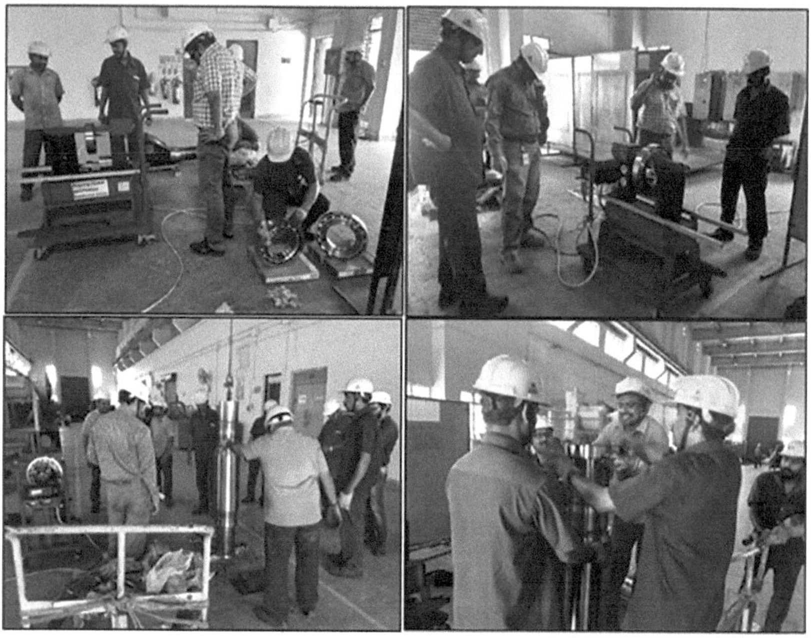

FIGURE 3.22 Installation of new bearings on fan shaft.

FIGURE 3.23 Fan blade installation.

All fan blades were installed on serviced rotor (see Figure 3.24). The alignment of an assembled rotor with motor and coupling was completed (see Figure 3.25). Fan boxed up (ready for operation) was done. Trial was taken at full load and all vibration readings were found well within limit (Figure 3.26).

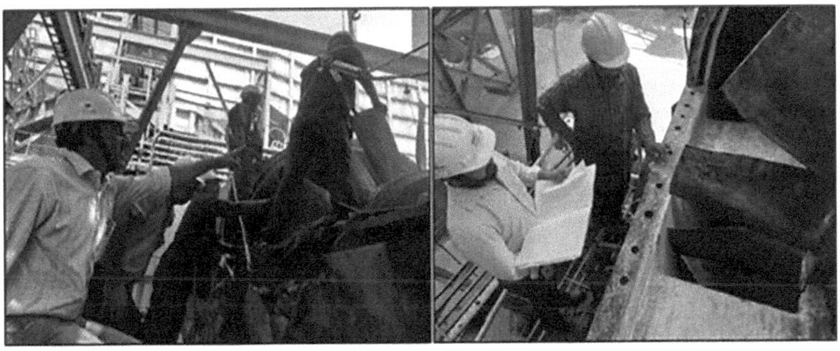

FIGURE 3.24 Fan blade clearance check.

FIGURE 3.25 Final alignment work.

FIGURE 3.26 Maintenance work completed.

3.9 POST-MAINTENANCE VIBRATION READINGS – OFFLINE AND ONLINE

Table 3.1 represents vibration data comparison for pre- and post-maintenance activity. The motor (offline) and fan casing (online and offline) horizontal vibration readings were reduced to normal level after replacing fan rotor bearings. Also, the crest factor was reduced to normal level.

TABLE 3.1
Pre- and Post-Maintenance Vibration Data

Vibration Report of Booster Fan – 01				
			Initial Readings	**After Replacing Fan Bearings**
			31/03/2020	**09/04/2020**
Point		**Position**	**Velocity (mm/s) RMS**	
Motor	(NDE)	H	**1.9**	**0.5**
		V	0.5	0.3
		A	0.6	0.4
	Acceleration - g	H	0.7	0.8
Motor	(DE)	H	**2.2**	**0.5**
		V	0.5	0.4
		A	0.7	0.5
	Acceleration - g	H	2.7	2.6
Fan casing – offline		H	**4.2**	**1.7**
Fan casing – online		H	**5.4**	**2.8**
Fan casing – crest factor		H	**4.7**	**3.5**

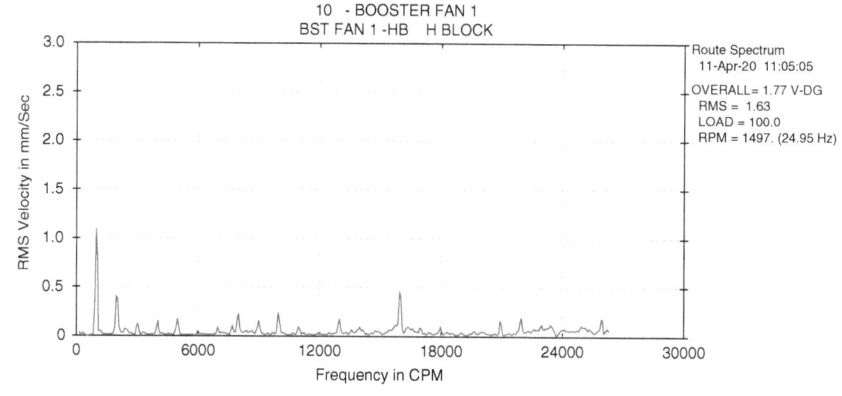

FIGURE 3.27 Spectrum of fan casing (Horizontal).

3.10 POST-MAINTENANCE: FFT SPECTRUM OF VIBRATION READINGS

The vibration FFT Spectrum shown in Figure 3.27 does not represent any abnormality such as bearing damage, blade pass frequency or looseness. It shows only normal frequencies of running speed 1× and 2× with low amplitude. Hence, this confirms the success of the maintenance work that was executed.

3.11 CONCLUSIONS

Booster fan is critical equipment that performs the function of carrying and transporting of flue gas from FGD scrubber to atmosphere via chimney (stack). This booster fan is provided with blade pitch control for controlling fan loading, inlet and outlet dampers for isolation to facilitate scrubber and fan maintenance. Since the booster fan has no standby, the reliability of fan needs to be very high. It is a critical auxiliary in the process, its failure will directly affect the tripping of FGD plant. In addition, if fan fails during operation and not available for a prolong period, it may lead to environmental deviation and may require unit outage, thereby generation loss happens in extreme cases. Outage of unit may hamper interruption in power supply to esteemed customers of commercial capital of India.

The present case study has clearly demonstrated that condition-based vibration monitoring is a vital component in rotating machinery analysis and detection of abnormality at an early stage. Vibration monitoring with FFT and time waveform spectrum analysis helped in diagnosing the exact root cause of abnormity well in advance avoiding secondary damage to fan and motor. Circle plot analysis and consideration of crest factor played a major supportive role in confirming the deterioration of bearings. Time waveform analysis is an important analysis tool. Presented data should be collected on every route and always be considered when attempting to diagnose and improve vibration analysis accuracy for machine faults which will help maintenance organization to become more proactive and accurate. The damage to motor and fan internals is avoided using these techniques and other major benefit includes prevention of downtime, avoidance of repair of major fan rotor or replacement costs and reliability of unit. The application of coordinated condition monitoring significantly helped in diagnosing the exact root cause of abnormity well in advance.

4 Case Study 3
Coal Mill 1CD: Gearbox Intermediate Gear Damage – September 2019

EXECUTIVE SUMMARY

The present case study is a success story of the implementation of condition monitoring techniques at ADTPS, wherein intermediate gear spacer looseness of main reducer gearbox of HT auxiliary Coal Mill – 1CD was diagnosed. Noise monitoring and vibration (FFT and Time waveform) spectrum analysis helped in diagnosing the exact root cause of abnormity well in advance. This has saved the plant from huge loss in terms of generation, unplanned breakdown time and associated maintenance cost.

On 13 September 2019, local operator noticed minor abnormal sound from main reducer gearbox during routine monitoring. Noise data were collected and recorded as 93.1 decibels (dB). No deviation was noticed in offline vibration and acceleration values. On 16 September 2019, noise level deviated from 93.1 to 102.8 dB. In addition, minor deviation noticed in acceleration value (1.6–2.3 g) of main reducer gearbox input NDE bearing. No further deviations were noticed in any process parameters and gearbox vibrations. It was decided to keep under close monitoring.

On 24 September 2019, noise level and acceleration value were increased to 107.1 decibels (dB) and 3.1 g, respectively. In addition, axial vibration of main reducer gearbox bearing was maintaining 12.0 mm/s. This issue was discussed in departmental meeting. Spectrum analysis revealed that there is increase in vibration of main reducer intermediate NDE bearing with high noise and found 414 cpm sidebands having gear mesh frequency 25851 cpm. The 414 rpm co-relates with intermediate gear speed and suspected issue of improper gear meshing. This initiated for gearbox internal inspection and coal mill was taken out of service and standby coal mill was taken in service to avoid any power generation loss.

On 25 September 2019, main reducer gearbox internal inspection was carried out, and loose intermediate gear spacer was found. Intermediate gear replaced with spare gear. Detailed inspection of input and output gear along with bearings was carried out in the presence of original equipment manufacturer (OEM). No abnormality was noticed. On 26 September 2019, the coal mill was taken in service at full load after replacing the main reducer intermediate gear. Gearbox noise level, vibration and acceleration value were reduced to 91.7 decibels (dB), 6.5 mm/s and 0.8 g, respectively.

4.1 PRE-MAINTENANCE HISTORY

On 13 September 2019, during routine monitoring, the local operator noticed minor abnormal sound from main reducer gearbox. As shown in Figure 4.1, the mechanical and

DOI: 10.1201/9781003432616-4

FIGURE 4.1 Verification of abnormal sound from gearbox.

FIGURE 4.2 Noise and ultrasound data collection.

condition monitoring team checked the intensity of this sound. Noise data was collected and recorded as 93.1 dB, and no deviation was noticed in ultrasonic reading (as shown in Figure 4.2). No deviation was noticed in offline vibration and acceleration values. On 16 September 2019, it was found that noise level deviated from 93.1 to 102.8 dB. This issue was discussed in daily HoD meeting and decided to keep under close monitoring. On 24 September 2019, noise level and acceleration value further increased to 107.1 dB and 3.1 g, respectively as shown in Figure 4.3.

It was observed that the axial vibration of main reducer gearbox bearing was increased to 12.0 mm/s. By observing such increasing trend of noise, vibration and acceleration value, it was decided to change over the coal mill and go for internal inspection of main reducer gearbox. In this regard SAP work order (300000115559) was raised to mechanical maintenance department and detailed diagnosis analysis was carried out.

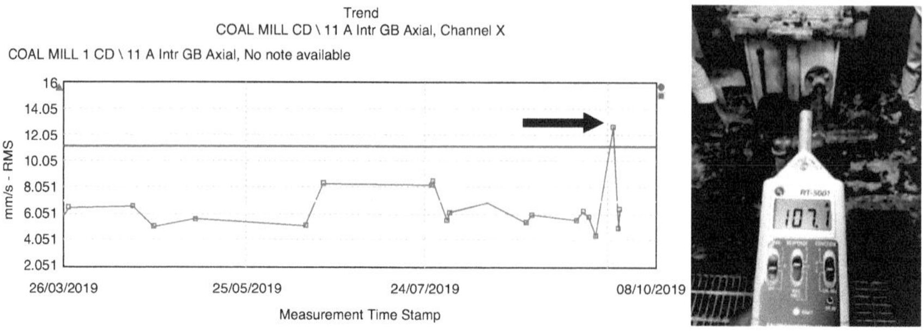

FIGURE 4.3 Acceleration value trend (LHS) and noise level (RHS).

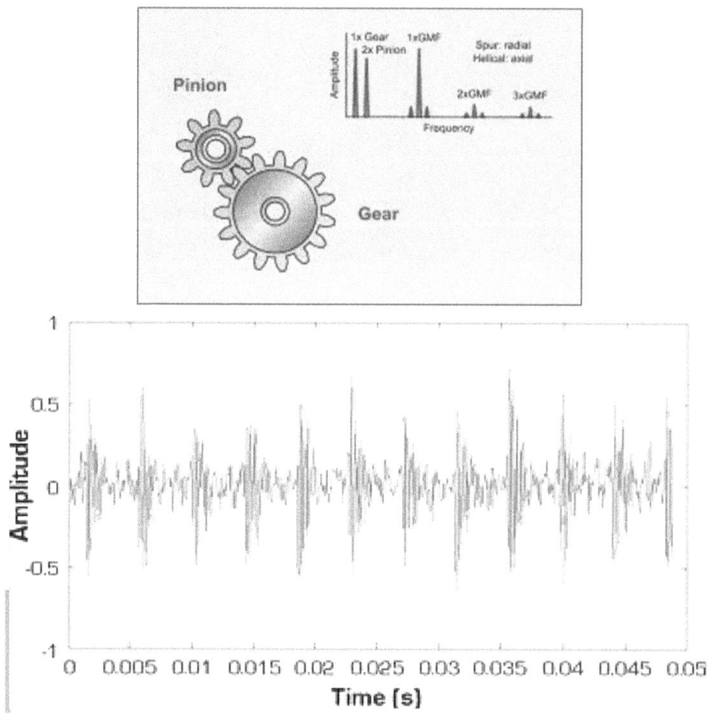

FIGURE 4.4 Fundamentals of time waveform.

4.2 TIME WAVEFORM - BASICS

The featured time waveform plot shows how the signal from sensor or probe appears when graphed as amplitude over time means amplitude versus time. The time waveform is a record of exactly what happened from moment to moment as the shaft turns and gears mesh (see Figure 4.4).

Changes in each second data which results from impacts, rubs, scrapes, rattles, surges, etc. is recorded in the time waveform and then summarized in the spectrum. Therefore, it is critical to record the time waveform correctly and analyze it when any suspicion fault condition exists. Time waveform analysis is especially useful for analyzing the gearboxes and bearings.

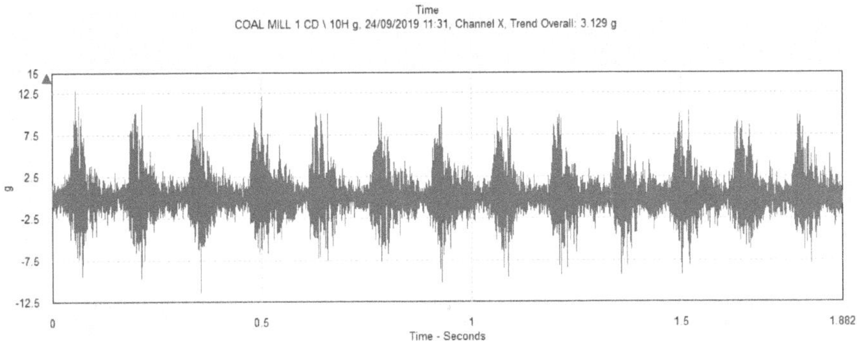

FIGURE 4.5 Time waveform of main reducer intermediate NDE bearing (horizontal).

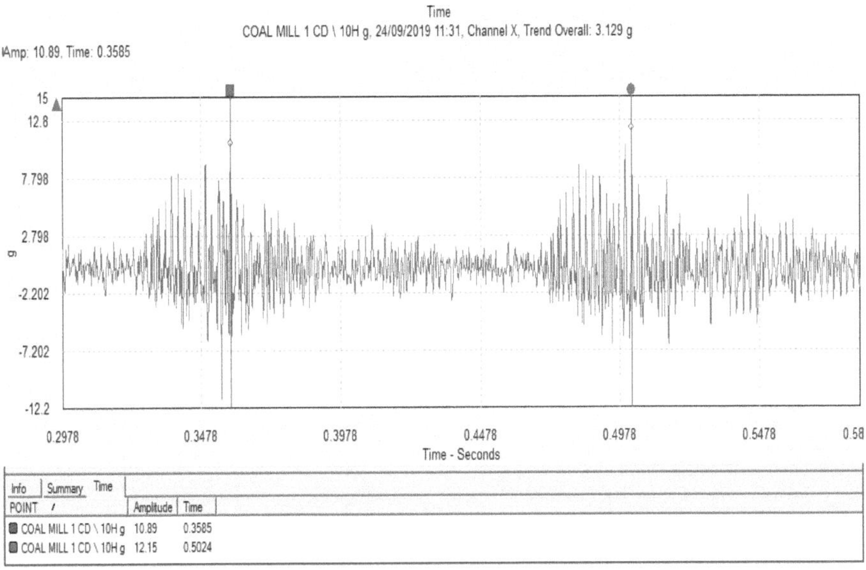

FIGURE 4.6 Zoomed waveform showing time between any two impacts.

4.3 TIME WAVEFORM ANALYSIS

Following is the time waveform spectrum collected from main reducer intermediate NDE bearing for finding the root cause of abnormality of abnormal sound and high vibration level (Figures 4.5 and 4.6).

Time difference between the two impacts was calculated as:

$T = 0.5024 - 0.3585 = 0.1439$ second

Frequency in Hz, $F = 1/T = 1/0.1439 = 6.9$ Hz

Frequency in CPM, $F = 6.9 \times 60 = 414$ cpm (co-relates with intermediate gear speed)

4.4 VIBRATION FFT SPECTRUM DIAGNOSIS

To verify the results obtained from time waveform analysis, acceleration spectrum was collected and analyzed (Figures 4.7 and 4.8).

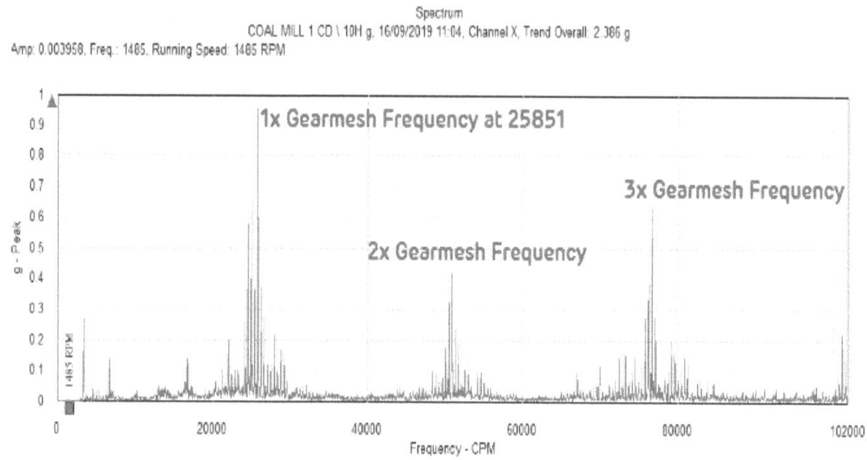

FIGURE 4.7 Acceleration spectrum of main reducer intermediate NDE bearing (horizontal).

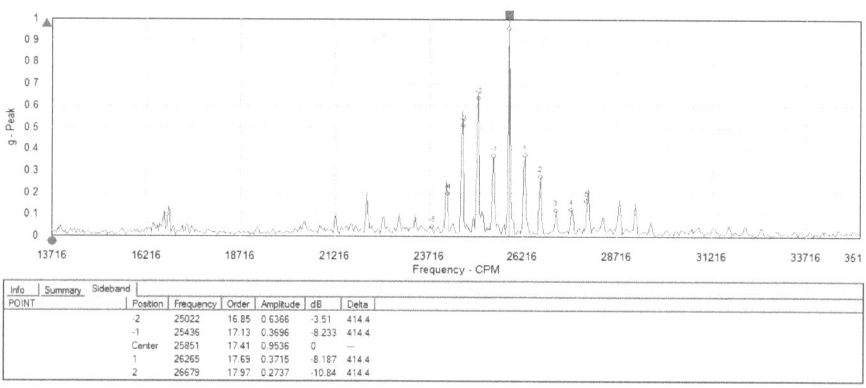

Position	Frequency	Order	Amplitude	dB	Delta
-2	25022	16.85	0.6366	-3.51	414.4
-1	25436	17.13	0.3696	-8.233	414.4
Center	25851	17.41	0.9536	0	---
1	26265	17.69	0.3715	-8.187	414.4
2	26679	17.97	0.2737	-10.84	414.4

FIGURE 4.8 Zoomed spectrum showing sidebands of 414 cpm around gear mesh frequency 25851 cpm.

It was found that there was an increase in vibration of main reducer intermediate NDE bearing with high noise. Spectrum analysis was carried out and found 414 cpm sidebands around 1× gear mesh frequency of 25851 cpm. Also, similar sidebands were noticed at 2× and 3× gear mesh frequency. For this main reducer gearbox, 414 rpm co-relates with intermediate gear speed and the suspected issue of improper gear meshing. This initiated for gearbox internal inspection for input and intermediate gears.

4.5 FREQUENCY ANALYSIS

Increase in noise level of main reducer gearbox was found 107.1 dB. Input NDE bearing acceleration deviation was also noticed from 1.6 to 3.1 g. Main reducer gearbox bearing axial vibration was maintained at 12.0 mm/s. Time waveform and FFT spectrum analysis revealed 414 cpm sidebands around 1× gear mesh frequency of 25851 cpm. Also, similar sidebands were noticed at 2× and 3× gear mesh frequency (Figure 4.9).

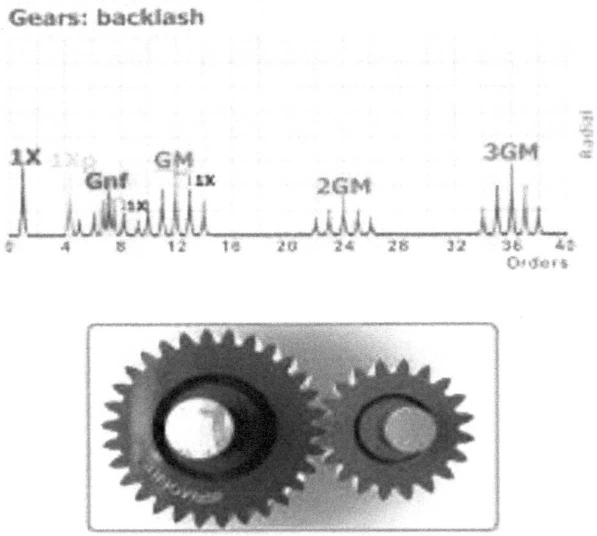

FIGURE 4.9 Fundamentals of gear backlash and gear mesh frequency.

For this main reducer gearbox, 414 rpm of sidebands co-relates with intermediate gear speed and suspected issue of improper gear meshing. This initiated for gearbox internal inspection of input and intermediate gears. It was concluded that main reducer gearbox bearing vibration, acceleration and noise level were increased due to internal impacts generated from loose spacer of intermediate gear. It was decided to changeover the coal mill for internal inspection of gearbox.

4.6 OBSERVATIONS DURING MAINTENANCE WORK

Condition of motor and main reducer gearbox side coupling was checked, and no deviation was noticed. Alignment between main reducer gearbox and pinion found disturbed by 0.4 mm radially. With reference to Figure 4.10, intermediate gear spacer was found loose at keyway which means the locking of wheel on shaft was found loose.

With reference to Figure 4.11, detailed inspection of input, intermediate and output bearings and gears was carried out in the presence of original equipment manufacturer (OEM) of gearbox. No significant deviations were observed. That means all the bearing clearances and gear backlash readings found well within limits.

4.7 PERFORMED MAINTENANCE WORK

Main reducer gearbox is de-coupled from the rest of the group drive as shown in Figure 4.12. Before dismantling, initial alignment readings were taken for reference and then the main reducer intermediate gear was taken out and replaced with spare gear as shown in Figure 4.13. Main reducer gearbox alignment was done with rest of drive. Coal mill was started at full load and all vibration, noise database was found well within limit.

FIGURE 4.10 Intermediate gear spacer (found loose at key-way).

FIGURE 4.11 Inspection of gearbox internals.

FIGURE 4.12 Decoupling of gearbox with rest of drive.

FIGURE 4.13 Replacement of intermediate gear with bearings.

4.8 PRE- AND POST-MAINTENANCE VIBRATION AND NOISE READINGS

Tables 4.1 and 4.2, and Figures 4.14 and 4.15 represent the comparison of vibration, acceleration, and noise readings for pre- and post-maintenance activities.

TABLE 4.1
Initial Vibration and Noise Readings Taken on 24 September 2019

Main Reducer Gearbox	Input DE	Input NDE	Intermediate NDE	Output DE
H (mm/s)	6.9	8.2	5.0	7.6
V (mm/s)	4.2	4.5	2.3	7.9
A (mm/s)	11.1	12.0	12.6	14.1
Acceleration (g)	2.1	3.1	2.1	2.2
Noise level (dB)	107.1			

TABLE 4.2
Final Vibration and Noise Readings Taken on 26 September 2019

Main Reducer Gearbox	Input DE	Input NDE	Intermediate NDE	Output DE
H (mm/s)	4.2	5.5	3.9	5.7
V (mm/s)	3.2	3.8	2.9	4.3
A (mm/s)	6.1	6.9	6.5	6.6
Acceleration (g)	0.8	1.2	0.8	0.7
Noise level (dB)	91.7			

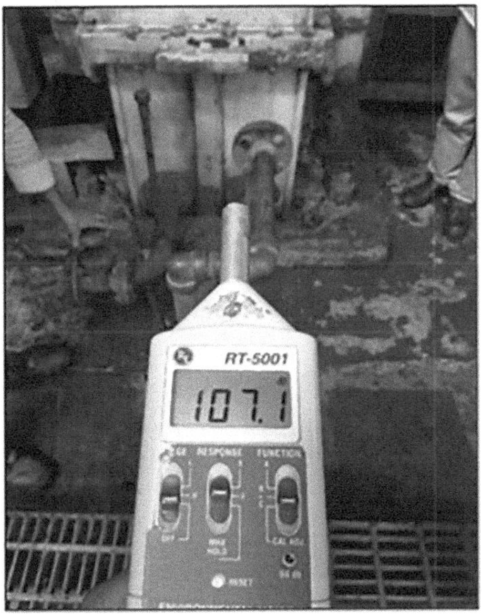

FIGURE 4.14 Pre-maintenance noise level.

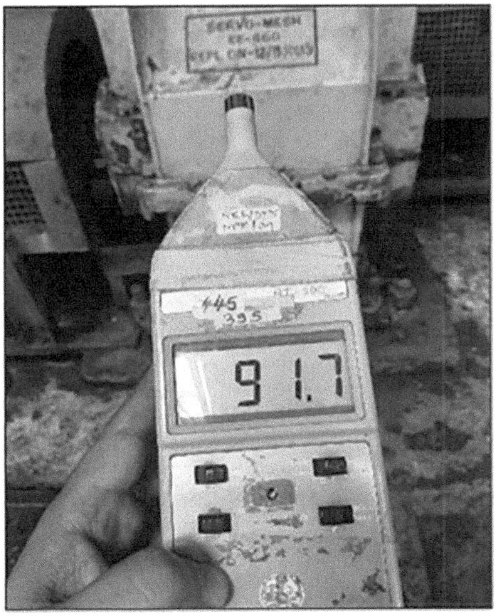

FIGURE 4.15 Post-maintenance noise level.

4.9 POST-MAINTENANCE TIME WAVEFORM AND FFT SPECTRUM

Figures 4.16 and 4.17 represent the post-maintenance time waveform and acceleration FFT spectrum to know the success of maintenance work execution. The spectrum does not show any sidebands of intermediate gear and hence can conclude the success of maintenance work executed.

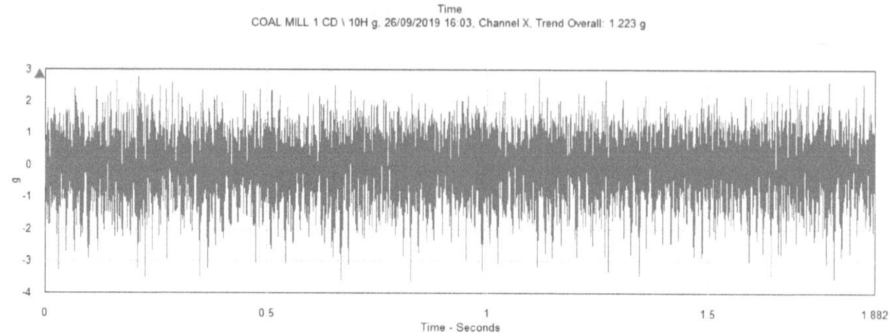

FIGURE 4.16 Acceleration time-waveform of reducer intermediate NDE bearing (horizontal).

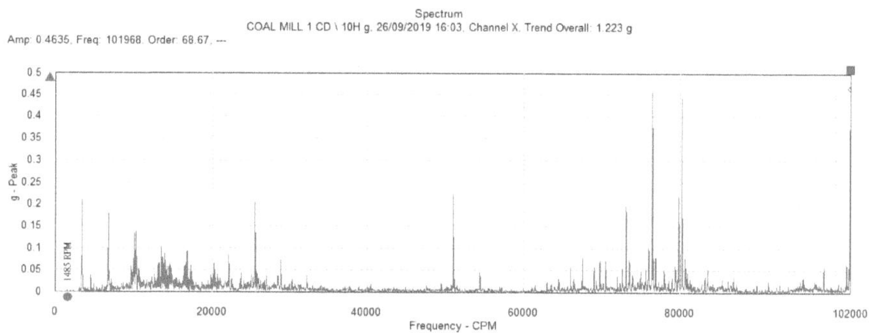

FIGURE 4.17 Acceleration FFT spectrum of reducer intermediate NDE bearing (horizontal).

4.10 CONCLUSIONS

Coal mill is an important auxiliary used to pulverize coal to required size for efficient combustion in boiler. Since coal mill supplies coal as fuel to boiler, the reliability needs to be very high. It is a critical auxiliary in the process because failure of it directly affects to the generation loss of 0.125 MUs/h which costs revenue of around Rs. 437,500 per hour. There is total three nos. of coal mills of 50% capacity at ADTPS (i.e. two running at full load and other is in stand by condition). This provision is made because coal mill is a critical equipment in the system and if it fails during non- availability of stand by mill, it may lead to tripping of entire plant and interruption in power supply to esteemed customers of Mumbai, India.

This case study has clearly demonstrated that noise monitoring is a vital tool for the detection of abnormality in rotating machinery at early stage. Time Waveform Analysis (TWA) played an important supportive role to initiate gearbox internal inspection and identifying the internal impacts generated from loose spacer of intermediate gear. Time waveform analysis is an important analysis tool. Time waveform data should be collected on every route and always be considered when attempting to diagnose and improve vibration analysis accuracy for machine faults that will help maintenance organization to become more proactive and accurate.

The damage of input, intermediate and output bearing, and gears is avoided using these techniques and other major benefit includes prevention of downtime, avoidancc of major gearbox repair or replacement costs and reliability of unit. The application of coordinated condition monitoring helped in diagnosing the exact root cause of abnormity well in advance.

5 Case Study 4
Coal Mill 1EF: Gearbox Input NDE Bearing Damage – September 2019

The present case study is a success story of the implementation of condition monitoring techniques at ADTPS, wherein Input NDE bearing deterioration of main reducer gearbox of HT auxiliary Coal Mill – 1EF was diagnosed. Noise monitoring and vibration spectrum analysis helped in diagnosing the exact root cause of abnormity well in advance. This has saved the plant from huge loss in terms of generation, unplanned breakdown time and associated maintenance cost.

On 26 August 2019, local operator noticed abnormal sound from main reducer gearbox during routine monitoring. Noise data was collected and found value on higher side i.e. 102.3 decibels (dB). Noise level compared with other running mills and found within limit of 94 decibels (dB). Based on the offline vibration database, it was observed that acceleration value of main reducer gearbox input NDE bearing was deviated from 2.5 to 6.8 g. No deviations were noticed in any process parameters and gearbox vibrations. It was decided to keep under close monitoring.

On 26 August 2019, noise level and acceleration value were increased to 103.6 decibels (dB) and 7.0 g, respectively. Acceleration spectrum analysis revealed that there is an increase in higher frequencies and the contribution of non-synchronous energy (any vibration that is greater than 1× frequency) is at higher side which confirms bearing defects. Observing the increasing trend of noise and acceleration value, it was decided to change over coal mill and to replace input NDE bearing of main reducer.

On 16 September 2019, coal mill was taken in service at full load after replacing Input NDE bearing of main reducer. Bearing noise level and acceleration value were reduced to 96.6 decibels (dB) and 2.0 g respectively. During maintenance, scoring marks were observed on bearing outer race and the same was replaced. Thorough maintenance work, inspection of input, intermediate and output bearings, and gears were carried out and no secondary damage/scoring marks were noticed.

5.1 PRE-MAINTENANCE HISTORY

On 26 August 2019, during routine monitoring, local operator noticed abnormal sound from main reducer gearbox. Noise data was collected and found noise value on higher side (i.e., 102.3 dB). Noise level was compared with other running coal mills and found within limit of 94 dB. After collecting offline vibration data, it was observed that the acceleration value of main reducer gearbox input NDE (non-drive end) bearing was deviated from 2.5 to 6.8 g (shown in Figure 5.1).

DOI: 10.1201/9781003432616-5

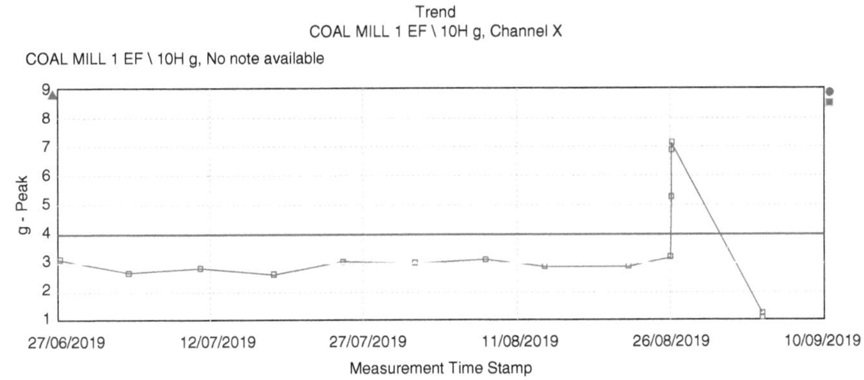

FIGURE 5.1 Acceleration value trend for gearbox input NDE bearing.

No deviations were noticed in any process parameters and gearbox vibrations. It was decided to keep under close monitoring. On 26 August 2019, noise level and acceleration value were increased to 103.6dB and 7.0g respectively. By observing such increasing trend of noise and acceleration value, it was decided to change over the coal mill to replace main reducer input NDE (non-drive end) bearing. For resolving this issue, work order (3000001114705) was raised in SAP software to mechanical maintenance department so that the mechanical team can execute the work after getting valid permit from operation department.

5.2 VIBRATION DIAGNOSIS

To find out of root cause, acceleration spectrum was collected and analysed (see Figure 5.2). The spectrum revealed that increase in higher frequencies and contribution of non-synchronous energy is at higher side which confirms the presence of bearing defects (see Figure 5.3).

Acceleration spectrum analysis revealed that there was increase in higher frequencies and with reference to Figure 5.3 the contribution of non-synchronous energy (any vibration that is greater than 1× frequency but is not a whole number multiple of it) was found at higher side which confirmed defects in bearing. The high frequencies got excited. These high frequencies were observed due to the bearing defect frequencies. Thus, it was confirmed that main reducer input NDE bearing deterioration was in progress.

5.3 FREQUENCY ANALYSIS

Noise level of main reducer gearbox was increased and recorded as 103.6dB. In addition, the deviation was noticed in acceleration value of input NDE bearing from 2.5 to 6.8g. Acceleration spectrum analysis revealed that increase in higher frequencies and contribution of non-synchronous energy was found at higher side which confirmed bearing defects (Figures 5.4 and 5.5).

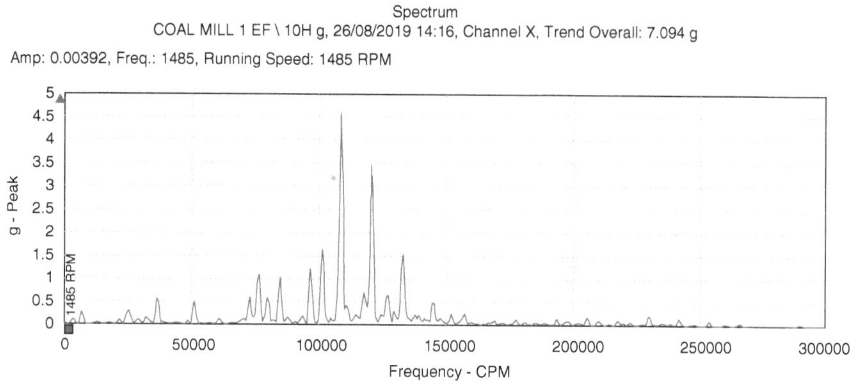

FIGURE 5.2 Acceleration spectrum of main reducer input NDE bearing (horizontal).

Info | Summary | Spectrum

POINT	Information	Value
	Threshold	0.05
	Spectrum Overall	7.095
	Sync	4.773
	Subsync	0
	Nonsync	5.25

FIGURE 5.3 Contribution of non-synchronous energy.

Synchronous	Non-Synchronous	Sub-Synchronous
• Mass imbalance • Looseness • Worn or damaged gears • Eccentric rotation • Misalignment • Reciprocating motion • Vane-pass	• Bearing defects • Cavitation • Electrical faults • Nearby machines and components • System resonances	• Nearby machines and components • Severe looseness • Rubbing of rotors or shafts • Belt vibrations

FIGURE 5.4 Causes of high non-synchronous energy.

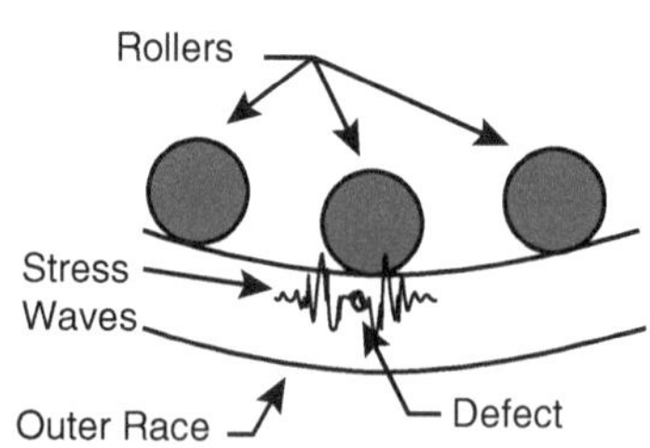

FIGURE 5.5 Bearing defect initiation.

These high-frequency stress waves' events occurred at repeatable time intervals corresponding to bearing defect frequency, then it is likely that bearing defect was present. It was concluded that acceleration and noise level of input NDE bearing were increased due to early-stage deterioration of bearing. It was decided to changeover the coal mill for replacement of gearbox input NDE bearing.

5.4 OBSERVATIONS DURING MAINTENANCE WORK

Condition of motor and main reducer gearbox side coupling was checked. No deviation was noticed. Alignment between main motor and main reducer gearbox was found well. Input NDE bearing clearances was found on higher side and scoring was observed at outer race as shown in Figure 5.6.

Detailed inspection of input, intermediate and output bearings, and all gears was carried out (as shown in Figure 5.7). All bearing clearances and gear backlash readings were found within limits and no deviations were recorded.

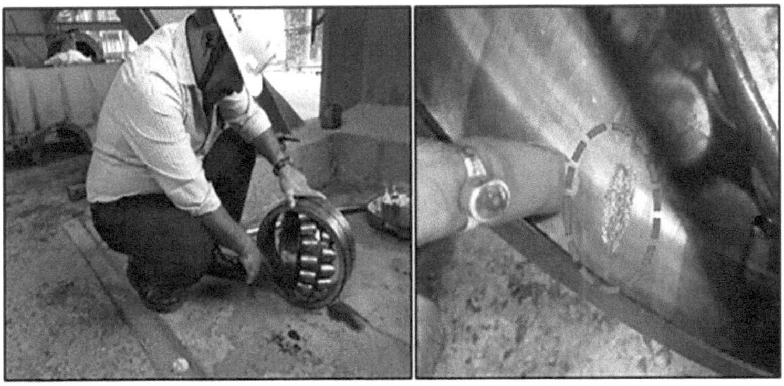

FIGURE 5.6 Scoring mark noticed on bearing outer race area.

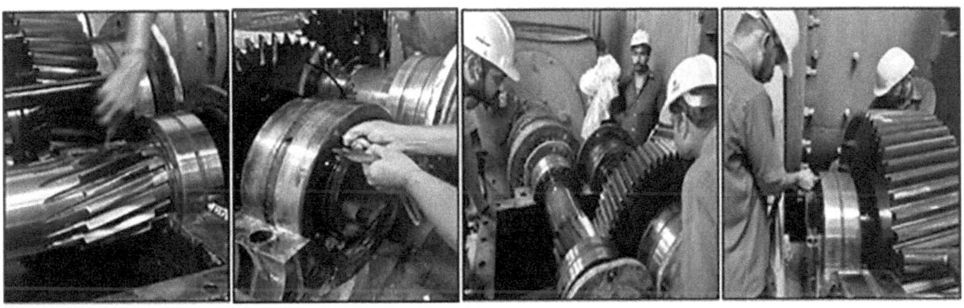

FIGURE 5.7 Internal inspection of gearbox.

5.5 PERFORMED MAINTENANCE WORK

Main reducer gearbox was de-coupled from the rest of the group drive. Before dismantling, initial alignment readings were taken for reference. Main reducer input shaft was taken out for removing the NDE (Non-drive end) bearing (shown in Figure 5.8).

Main reducer input NDE bearing was removed, and new bearing was installed. Main reducer input shaft with new NDE bearing put back in gearbox. Main reducer gearbox alignment was done with the rest of drive. Coal mill was taken in trail at full load and all vibration, noise data were found well within limit (Figure 5.9).

FIGURE 5.8 Removal of main reducer input shaft.

FIGURE 5.9 Main reducer input NDE bearing replacement work.

5.6 PRE- AND POST-MAINTENANCE VIBRATION AND NOISE READINGS

Tables 5.1 and 5.2 and Figures 5.10 and 5.11 represent the comparison of vibration, acceleration and noise readings for pre- and post-maintenance activities.

TABLE 5.1
Initial Vibration and Noise Readings Taken on 26 August 2019

Main Reducer Gearbox	Input DE	Input NDE	Intermediate NDE	Output DE
H (mm/s)	3.3	4.1	3.1	3.2
V (mm/s)	2.7	4.8	2.9	3.6
A (mm/s)	5.3	6.0	5.3	5.9
Acceleration (g)	2.2	7.0	5.1	1.7
Noise level (dB)	103.6			

TABLE 5.2
Final Vibration and Noise Readings Taken on 16 September 2019

Main Reducer Gearbox	Input DE	Input NDE	Intermediate NDE	Output DE
H (mm/s)	3.9	4.1	3.8	3.7
V (mm/s)	2.8	4.8	2.4	3.9
A (mm/s)	3.6	3.9	3.3	4.1
Acceleration (g)	1.2	2.0	1.3	1.3
Noise level (dB)	96.6			

FIGURE 5.10 Pre-maintenance noise level.

FIGURE 5.11 Post-maintenance noise level.

5.7 CONCLUSIONS

Coal mill is an important auxiliary used to pulverize the coal into required size for efficient combustion in boiler. Since coal mill supplies coal as fuel to boiler, the reliability needs to be very high. It is a critical auxiliary in the process because failure of it directly affects the generation loss of 0.125 MUs/h which costs revenue of around Rs. 437,500 per hour. There are three coal mills of 50% capacity at ADTPS (i.e., two running at full load and the other is in stand-by condition). This provision is made because coal mill is a critical equipment in the system and if it fails during the non-availability of stand by mill, it may lead to tripping of plant and interruption in power supply to esteemed customers of Mumbai, India.

This case study has clearly demonstrated that noise monitoring is a vital tool for the detection of abnormality in rotating machinery at an early stage. Deviation in acceleration value played an important supportive role to confirm the deterioration of gearbox input NDE bearing even before any noticeable changes are reflected in vibration and any other process parameters. Each vibration analyst should use this tool to improve vibration analysis accuracy which can help maintenance organization to become more proactive and accurate.

The damage of intermediate and output bearing, and gears is avoided by using these techniques and other major benefit includes prevention of downtime, avoidance of major gearbox repair or replacement costs and reliability of unit. The application of coordinated condition monitoring helps in diagnosing the exact root cause of abnormity well in advance.

6 Case Study 5 *Coal Mill 2CD: Output Gear Shaft Crack Detection – July 2019*

EXECUTIVE SUMMARY

The present case study is a success story of the implementation of oil condition monitoring program at ADTPS, wherein output gear shaft crack of main reducer gearbox of HT auxiliary Coal Mill – 2CD was detected. Repetitive high wear particle concentration of main reducer gearbox, oil analysis initiated a visual inspection and suspected that despite changing intermediate gear, bearings, and oil during maintenance work, still wear particle concentration count was maintained on higher side (probably due to crack), which helped in diagnosing the root cause of abnormity in advance and timely action has avoided the major gearbox repair.

On 12 October 2018, wear debris analysis report stated the high wear particle concentration (911 WPC) with wear particles (gear) of size ranging up to 32 μm and wear particles (bearing) of size ranging up to 72 μm which indicated excessive load on bearings. On 13 October 2018, oil parameter deviation was discussed in departmental meeting and decided to inspect the bearing of main reducer gearbox during the upcoming maintenance opportunity.

On 20 November 2018, bearing of main reducer gearbox and intermediate gear replaced with a spare gear. Fresh oil top-up was done (420 L). On 10 December 2018, wear debris analysis report stated high wear particle concentration (814 WPC) with wear particles (gear) of size ranging up to 26 μm and wear particles (bearing) of size ranging up to 64 μm. On 11.12.2018, high wear particle concentration issue was discussed in departmental meeting and decided to resample after 15 days with close monitoring of vibration and acceleration parameters.

On 6 March 2019, wear debris analysis report stated high wear particle concentration (1089 WPC) against the limit value (1000 WPC) with wear particles (gear) of size ranging up to 24 μm and wear particles (bearing) of size ranging up to 68 μm. On 8 March 2019, detailed oil analysis report was discussed in departmental meeting and decided to go for internal inspection of main reducer gearbox during upcoming maintenance opportunity.

On 16 April 2019, the main reducer gearbox internal inspection was carried out and the crack was noticed on output gear shaft. Output gear was immediately replaced with spare gear. Detailed inspection of input and intermediate gear and bearings were carried out and no abnormality was noticed. On 17 April 2019, the coal mill was taken in service after replacing the output gear. All vibration readings were found normal. On 4 July 2019, the wear debris analysis report stated reduction in wear particles (gear) concentration was from 1089 to 868 and the reduction of wear particles (bearing) was from 68 to 36 μm.

This case study clearly demonstrated that oil condition monitoring is a vital tool in rotating machinery analysis and detection of abnormality at early stage even before deviation to be reflected in vibration, acceleration, noise and any other process parameters.

DOI: 10.1201/9781003432616-6

6.1 PRE-MAINTENANCE HISTORY

On 12 October 2018, wear debris analysis report stated high wear particle concentration 911 WPC with gear wear particles of size ranging up to 32 μm and bearing wear particles of size ranging up to 72 μm which indicated excessive load on bearings.

On 13 October 2019, work order (300000097879) was raised in SAP software to mechanical maintenance department for internal inspection of main reducer gearbox. On 13 October 2018, oil parameter deviation was discussed in plant HoD meeting and decided to inspect main reducer gearbox bearings during opportunity. On 20 November 2018, main reducer gearbox bearings, intermediate gear were replaced with spare one in the presence of OEM (Original equipment manufacture of gearbox). Fresh oil top-up was done (420 L) as shown in Figure 6.2.

On 10 December 2018, wear debris analysis report stated high wear particle concentration of 814 WPC with gear wear particles of size ranging up to 26 μm and bearing wear particles of size ranging up to 64 μm (shown in Figure 6.3).

On 11 December 2018, high wear particle concentration issue was discussed in HoD meeting and decided to resample after 15 days with close monitoring of vibration and acceleration parameters. On 6 March 2019, wear debris analysis report stated high

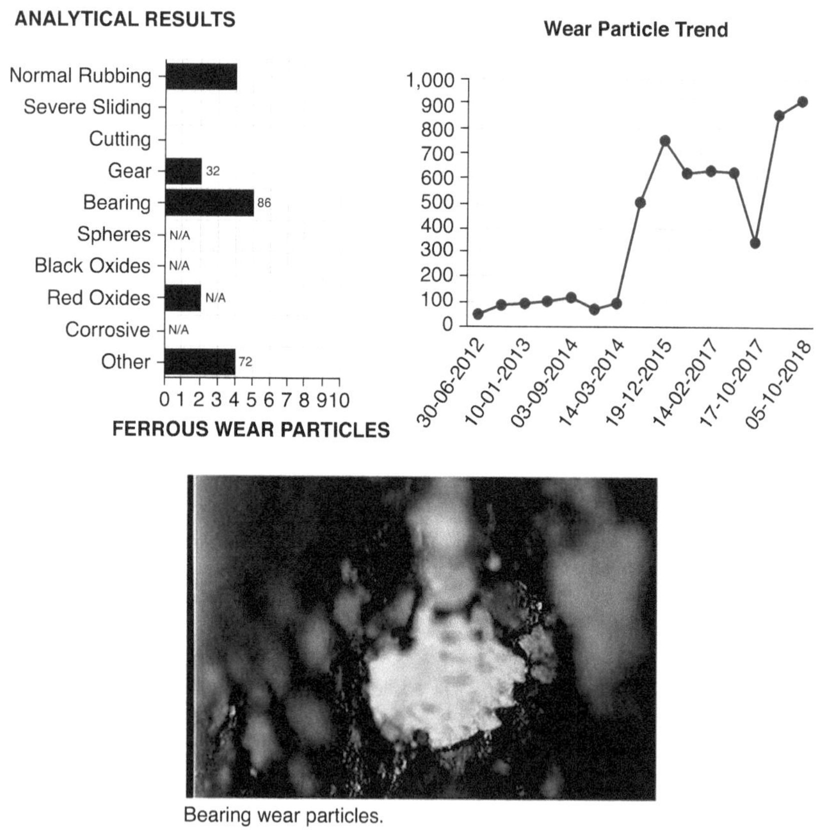

FIGURE 6.1 Wear particle concentration (WPC) trend for the report dated 12 October 2018.

FIGURE 6.2 Inspection of Intermediate gear (LHS) and oil top up (RHS).

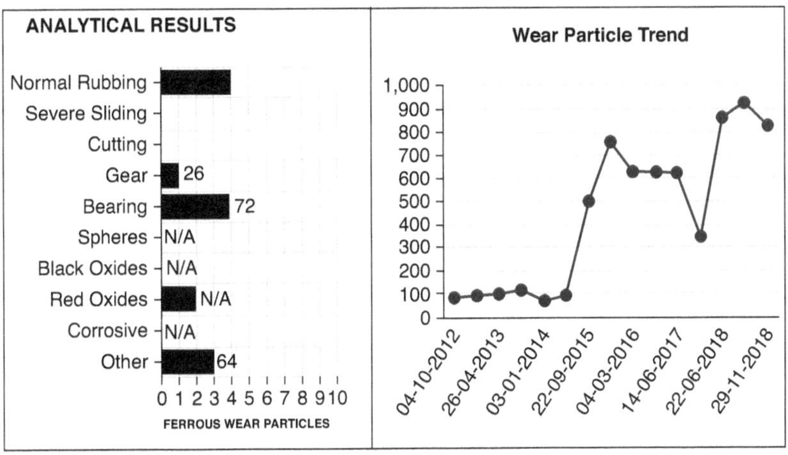

FIGURE 6.3 Wear particle concentration (WPC) trend for report dated 10 December 2018.

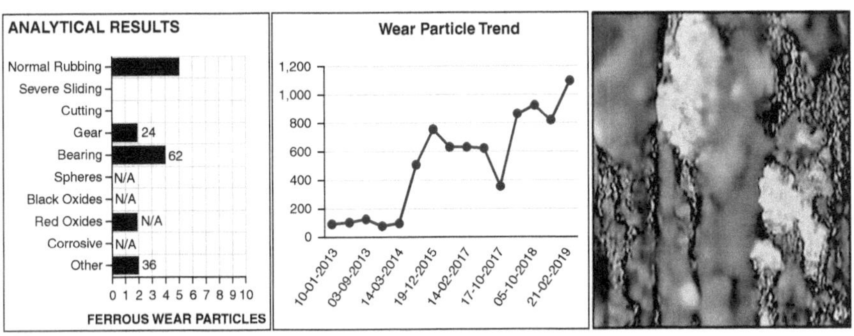

FIGURE 6.4 Wear particle concentration (WPC) trend for report dated 6 March 2019.

wear particle concentration of 1089 WPC against a limit value of 1000 WPC with gear wear particles of size ranging up to 24 μm and bearing wear particles of size ranging up to 68 μm (as shown in Figure 6.4).

No deviation was observed in vibration, acceleration, noise and any other process parameters. Repetitive oil parameter deviation for oil samples dated 12 October 2018, 10 December 2018 and 6 March 2019 of reducer gearbox oil initiated a visual inspection and suspected that in spite of changing intermediate gear, bearings and oil the WPC count is maintained on higher side (probably due to crack). On 8 March 2019, work order (300000105508) was raised on SAP software to mechanical maintenance department for internal inspection. Detailed oil analysis report was discussed in HoD meeting. It was decided to go for an internal inspection of main reducer gearbox during opportunity.

6.2 OBSERVATIONS DURING MAINTENANCE WORK

On 16 April 2019, internal inspection of main reducer gearbox input, intermediate and output gears was done as shown in Figure 6.5. No abnormality was noticed for any bearing clearances and gear backlash reading.

Output gear shaft was found cracked at DE side (originated from keyway) as shown in Figure 6.6.

LPT (Liquid penetration test) test was carried out for confirmation of shaft surface crack as shown in Figure 6.7.

6.3 PERFORMED MAINTENANCE WORK

Main reducer gearbox was decoupled from the rest of the group drive as shown in Figure 6.8. Main reducer gearbox top cover was removed, output gear shaft was

FIGURE 6.5 Internal inspection of main reducer gearbox.

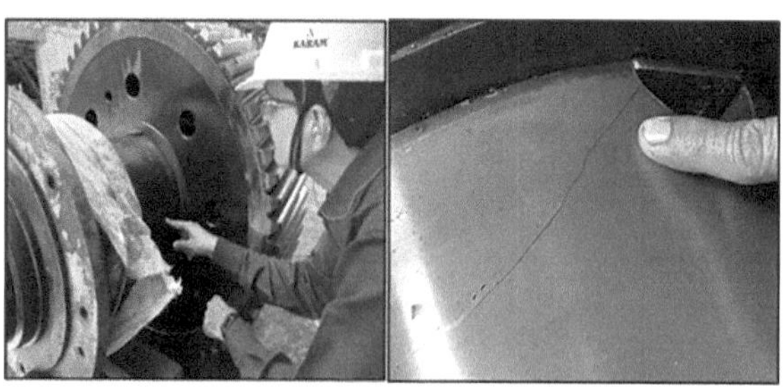

FIGURE 6.6 Crack observed on output gear shaft.

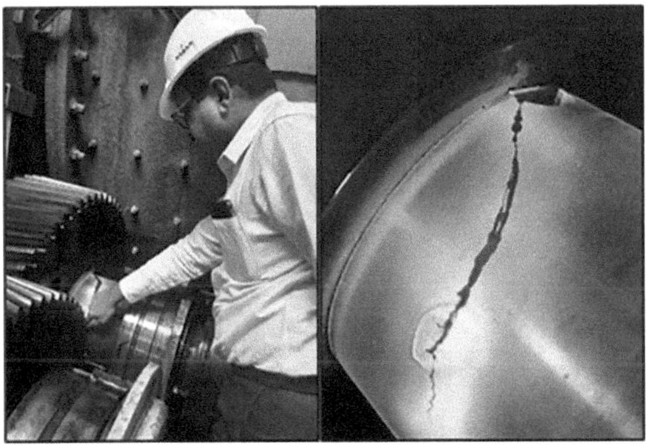

FIGURE 6.7 LPT test performed for crack confirmation.

FIGURE 6.8 Decoupling of gearbox from rest of drive.

taken out and coupling was removed as shown in Figure 6.9. The same coupling was installed on new spare output gear shaft. New output gear was installed in gearbox and top cover was placed as shown in Figure 6.10. Main reducer gearbox was aligned with rest of drive. On 17 April 2019, coal mill was taken in trail, and all vibration readings were found normal. On 4 July 2019, wear debris analysis report stated reduction in wear particle concentration with reduction in bearing wear particle size.

6.4 POST-MAINTENANCE OIL ANALYSIS

After replacing the output shaft with a new one, On 4 July 2019, again oil sample was collected and sent for analysis. Wear debris analysis report stated reduction in wear particle concentration from 1089 to 868 WPC with reduction in bearing wear particles from 68 to 36 μm (see Figures 6.11 and 6.12).

FIGURE 6.9 Removal of coupling from cracked output gear shaft.

FIGURE 6.10 Installation of coupling on new output gear shaft.

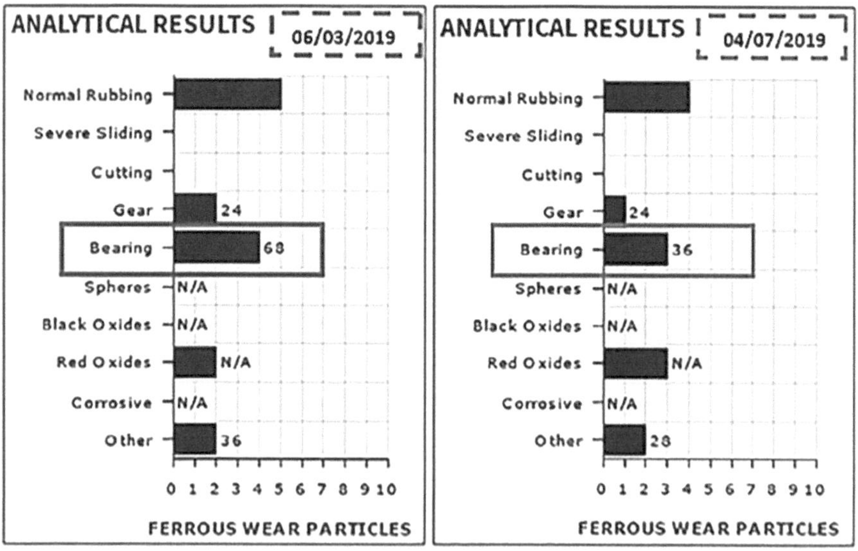

FIGURE 6.11 Pre- and post-maintenance oil analysis results.

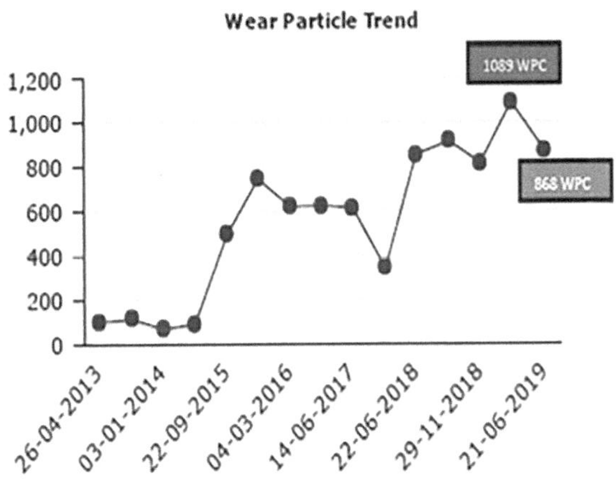

FIGURE 6.12 Wear particle concentration (WPC) trend for the report dated on 4 July 2019.

6.5 CONCLUSIONS

Coal Mill is an important auxiliary, used to pulverize the coal into required size for efficient combustion in boiler. Since coal mill supplies coal as fuel to boiler, the reliability needs to be very high. It is a critical auxiliary in the process because failure of it directly affects the generation loss of 0.125 MUs/h which costs revenue of around Rs. 4,37,500 per hour. There are three coal mills of 50% capacity (i.e., two running at full load and the other is in stand-by condition). This provision is made because the coal mill is critical equipment in the system, and if it fails during the non-availability of stand-by mill, it may lead to tripping of plant and interruption in power supply to our valued customers of Mumbai.

This case study has clearly demonstrated that oil condition monitoring is a vital tool in rotating machinery analysis and detection of abnormality at an early stage even before deviation is reflected in vibration, acceleration, noise and any other process parameters. Repetitive high wear particle concentration (WPC) of reducer gearbox oil analysis initiated a visual inspection and suspected that in spite of changing intermediate gear, bearings and oil, WPC count was found on higher side.

This helped in diagnosing the root cause of abnormity well in advance which can avoid secondary damage to main reducer gearbox. This has saved the plant from huge losses in terms of unplanned breakdown time and associated maintenance, manpower and spare costs. Other major benefits include the prevention of downtime, avoidance of major gearbox repair or replacement costs and reliability of unit.

7 Case Study 6
Booster Fan 01 (FGD): Fan Blade Crack Detection – May 2019

EXECUTIVE SUMMARY

The present case study is a success story of the implementation of condition monitoring techniques at ADTPS, wherein Fan blade crack of HT auxiliary, Booster Fan 02 was diagnosed. Noise monitoring deviation initiated a visual inspection which helped in diagnosing the exact root cause of abnormity well in advance. This has saved the plant from huge losses in terms of unplanned breakdown and associated maintenance, manpower and cost.

On 1 May 2019, plant control room (PCR) desk engineer noticed deviation in noise reading of fan from 03 to 05 dB which further increased up to 30 dB gradually during the routine monitoring. Fan was tripped due to 'high noise level' protection. No deviation was observed in vibration and any other process parameters. It has been decided to go for internal inspection before starting the fan.

On 2 May 2019, fan top cover was removed, and internal inspection was carried out. The crack in blade no. 03 was found and tip of blade worn out. In addition, bottom casing liner countersunk bolts (02 nos.) were found dislocated. Detailed inspection of fan rotor, bearing housing, shaft and blades was carried out. No secondary damage/scoring marks were noticed.

On 4 May 2019, trial of fan was taken at full load after replacing fan blades No. 02 and 03. It is observed that fan noise reading reduced from 30 to 03 dB and fan casing vibration was found stable. This case study has clearly demonstrated that condition-based noise monitoring is a vital component in rotating machinery analysis and detection of abnormality at early stage.

7.1 PRE-MAINTENANCE HISTORY

On 1 May 2019, during routine monitoring, plant control room (PCR) desk engineer noticed deviation in online fan noise reading from 03 to 05 dB (refer to Figure 7.1) and after some time, noise level further increased to 30 dB gradually. Fan was tripped due to high noise level protection. No other deviation was observed in fan casing regarding to vibration and other process parameters. Noise monitoring deviation was initiated for visual internal inspection of fan. Accordingly, work order (300000107995) was raised on SAP software to mechanical maintenance department for internal inspection of fan.

 DOI: 10.1201/9781003432616-7

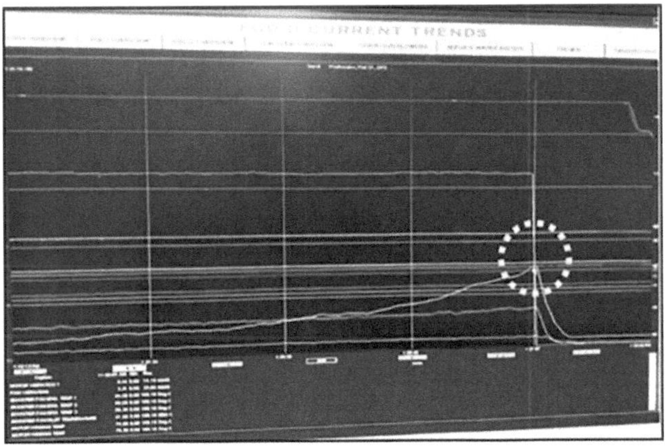

FIGURE 7.1 Online fan noise level trend.

7.2 OBSERVATIONS DURING MAINTENANCE WORK

Condition of motor and fan side coupling was checked. No deviation was noticed. Detailed inspection of fan rotor was carried out. No abnormality was observed. However, fan blade no. 03 was found cracked and tip of blade was worn out as shown in Figure 7.2. In addition, fan blade No. 02 profile was found disturbed (see Figure 7.2). Bottom casing liner countersunk bolts (02 nos.) were found dislocated, too.

7.3 PERFORMED MAINTENANCE WORK

Fan top cover was removed. All fan blades (16 nos.) were taken out. Overall inspection was carried out. Fan rotor and duct internal cleaning was done. Fan blade no. 02 and 03 were replaced with spare ones of the same weight (shown in Figure 7.3).

Liquid Penetration Test (LPT) was performed for all blades (16 nos.) and shaft for any surface crack identification as shown in Figure 7.4. No abnormality was noticed.

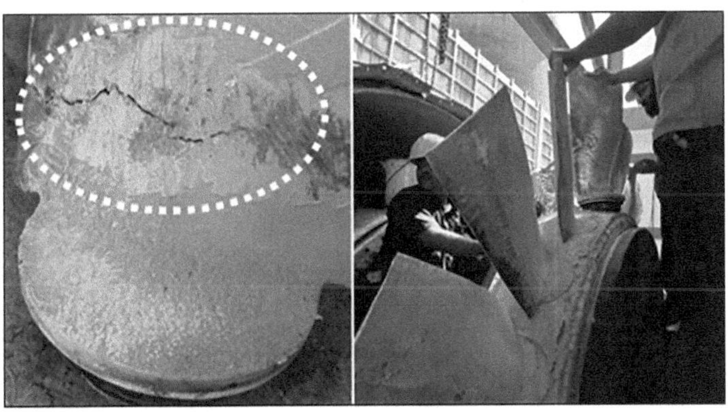

FIGURE 7.2 Cracked fan blade no. 02 and 03.

FIGURE 7.3 Replacement of fan blade no. 02 and 03.

FIGURE 7.4 LPT test done for all blades.

Ultrasonic test (UT) was performed for all blades and shaft for any internal crack identification as shown in Figure 7.5. No abnormality was noticed. Bottom casing liner countersunk bolts (02 nos.) were replaced with new ones and fan top cover was placed properly and then trial was taken at full load. All vibration and noise level data were found well within limit.

7.4 POST-MAINTENANCE VIBRATION AND NOISE READINGS

Table 7.1 shows pre- and post-maintenance vibration and noise data comparison readings to know the success of maintenance work execution.

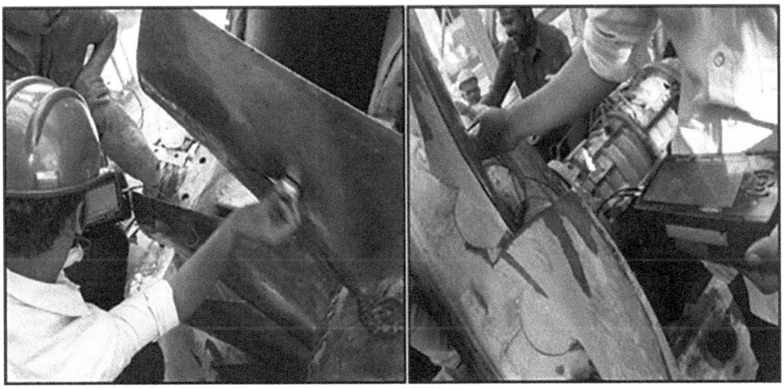

FIGURE 7.5 UT test done for all blades.

TABLE 7.1
Vibration and Noise Report of Booster Fan – 02

Point		Position	Initial Readings 01/05/2019	After Replacing Fan Blade No. 02 and 03 04/05/2019
			Velocity (mm/s) RMS	
Motor	(NDE)	H	0.5	0.5
		V	0.3	0.4
		A	0.9	0.5
	Acceleration - g	H	0.6	0.9
Motor	(DE)	H	0.8	0.6
		V	0.4	0.6
		A	0.5	0.6
	Acceleration - g	H	1.8	1.3
Fan casing – offline		H	1.8	2.1
Fan casing – online		H	2.9	3.2
Online fan noise level			**30 dB**	**03 dB**

7.5 CONCLUSIONS

Booster fan is critical equipment that carries and transports the flue gas from FGD scrubber to atmosphere via chimney. This booster fan is provided with blade pitch control for controlling the loading of fan and inlet and outlet dampers for isolation facility to execute booster fan maintenance.

Since booster fan has no standby, reliability of fan needs to be very high. However, it is a critical auxiliary in the process. If it fails, it will directly affect or trip the FGD plant. If its failure during operation and replacement is not available for a prolonged period, it may lead to environmental deviation. This requires unit outage, thereby generating loss in extreme cases. Outage of unit can hamper interruption in power supply to esteemed customers of Mumbai, India.

The present case study has clearly demonstrated that condition-based noise monitoring is a vital component in rotating machinery analysis and detection of abnormality at an early stage. Noise monitoring deviation initiated a visual inspection which helped in diagnosing the exact root cause of abnormity well in advance which avoid secondary damage to fan and motor. Other major benefits include the prevention of downtime, avoidance of major fan rotor repair or replacement costs and reliability of unit.

8 Case Study 7
PA Fan 2B: Fan Bearing Deterioration – April 2019

EXECUTIVE SUMMARY

The present case study is a success story of the implementation of condition monitoring techniques at ADTPS, wherein deterioration of fan bearing of HT auxiliary PA Fan – 2B was diagnosed. Vibration monitoring and spectrum analysis helped in diagnosing the exact root cause of abnormity well in advance.

On 22 February 2019, PCR desk engineer noticed deviation in fan casing vibration from 4.9 to 6.2 mm/s (RMS) during the routine monitoring. Based on the collected vibration database, it was observed that horizontal and vertical vibrations of fan casing were deviated from 3.5 to 4.0 and 2.9 to 3.2 mm/s (RMS) respectively. No other deviations were noticed in any process parameters and in motor vibrations. It was decided to keep under close monitoring.

On 15 March 2019, fan casing online vibration increased to 8.1 mm/s, and offline horizontal and vertical vibrations increased to 6.0 and 4.6 mm/s respectively. Spectrum analysis revealed that there was an increase in bearing inner race and ball spin defect frequency 17700 and 22230 cpm, respectively. These high frequencies lead to be the bearing defect frequencies. Thus, it was confirmed that fan bearing deterioration was in progress. This problem was discussed in the departmental meeting and decided to run the fan under close monitoring with a limit value of up to 11.0 mm/s (RMS).

On 1 April 2019, fan casing vibration increased to 11.2 mm/s. By observing such increment in fan casing vibration trend, it was decided to take shutdown of PA fan 2B in a planned way to replace fan rotor with a spare one. On 2 April 2019, fan trial was successfully taken at full load after replacing fan rotor. Fan casing vibration was reduced from 11.2 to 2.9 mm/s. During rotor maintenance, scoring marks were observed on fan bearing, and the same were replaced and kept ready as a standby rotor. During maintenance work, detailed inspection of fan rotor, bearing housing and shaft was carried out. No secondary damage/scoring marks were noticed.

This case study has clearly demonstrated that condition-based vibration monitoring is a vital component in rotating machinery analysis and detection of abnormality at an early stage.

8.1 PRE-MAINTENANCE HISTORY

On 22 February 2019, during routine monitoring, Plant Control Room (PCR) desk engineer noticed deviation in vibration reading of fan casing from 4.9 to 6.2 mm/s (RMS). By getting offline vibration data, it was observed that the horizontal and vertical vibration of fan casing deviated from 3.5 to 4.0 and 2.9 to 3.2 mm/s (RMS) respectively. No deviations were noticed in any process parameters and motor vibrations. It was decided to keep under close monitoring. Figure 8.1 shows all details of

DOI: 10.1201/9781003432616-8

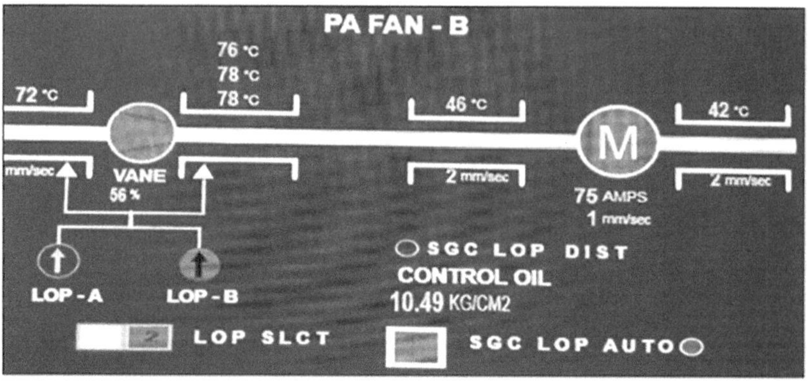

FIGURE 8.1 Fan process parameter readings.

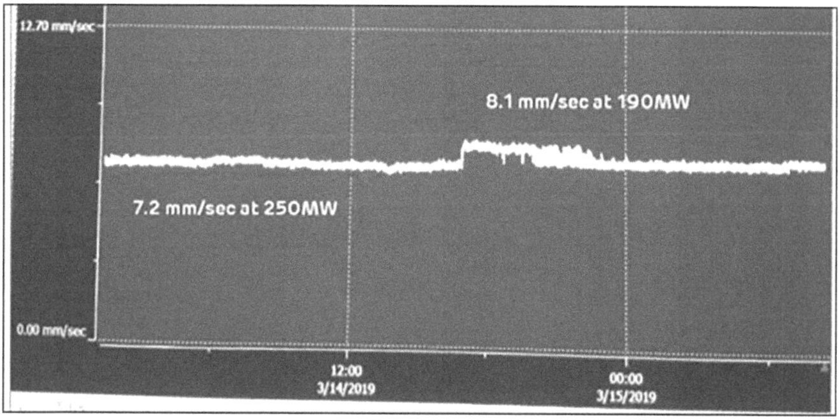

FIGURE 8.2 Online fan vibration trend.

process parameters of fan such as motor current, suction vane opening of fan and all bearing temperatures along with control oil pressure.

On 15 March 2019, online vibration of fan casing was increased to 8.1 mm/s as shown in Figure 8.2. Offline horizontal and vertical vibrations were observed to 6.0 and 4.6 mm/s, respectively as shown in Figure 8.3. The vibration level was found deviated than its normal value. High vibration issue was discussed in daily plant HoD meeting. It was decided to run fan under close monitoring with online limit value upto 11.0 mm/s (RMS). On 30 March 2019, online vibration of fan casing was found to be stable as 9.87 mm/s (see Figure 8.4).

On 1 April 2019, the online vibration of fan casing was increased to 11.2 mm/s as shown in Figure 8.5. Meanwhile noticed that fan casing offline horizontal and vertical vibration level also increased to 8.1 and 5.8 mm/s respectively.

Observing increasing trend of fan casing vibration (see Figure 8.6), it was decided to take the planned outage of fan to replace fan rotor with spare one and work order (300000105828) was raised on SAP software to mechanical maintenance department.

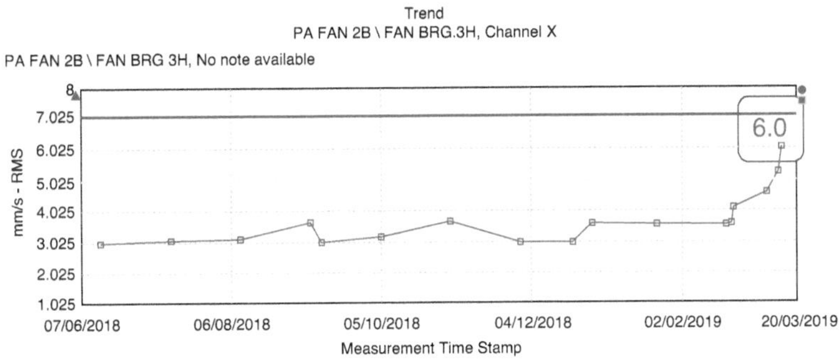

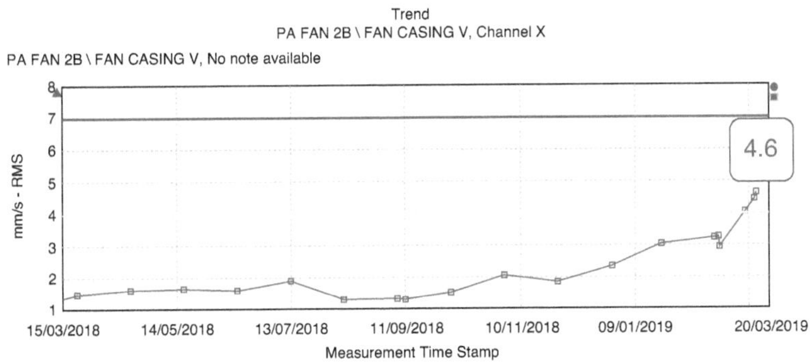

FIGURE 8.3 Offline fan casing vibration trend (horizontal and vertical).

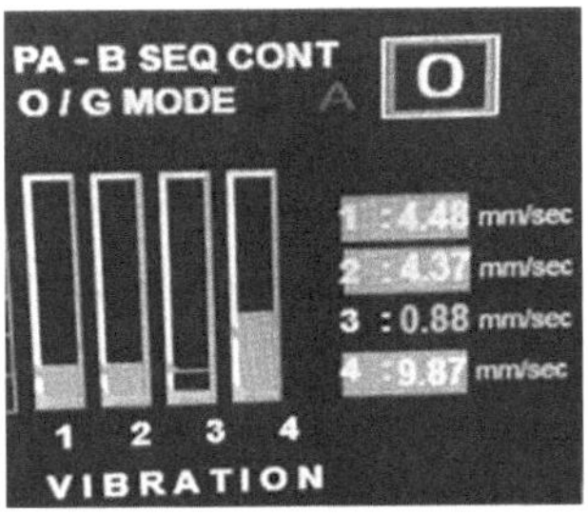

FIGURE 8.4 Online motor and fan vibration reading recorded on 30 March 2019.

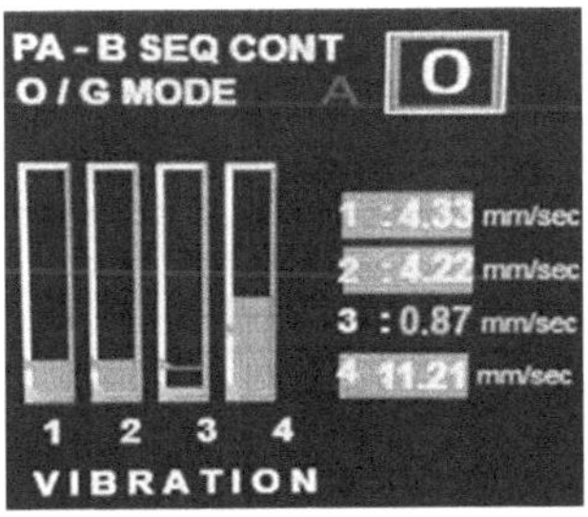

FIGURE 8.5 Online motor and fan vibration reading recorded on 1 April 2019.

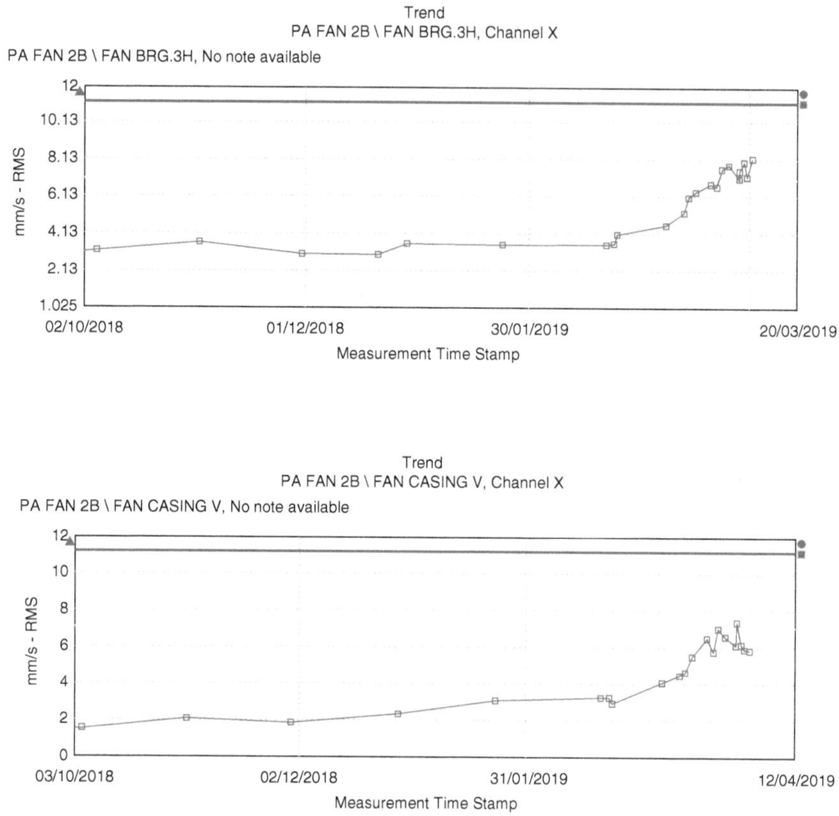

FIGURE 8.6 Offline fan casing vibration trend (horizontal and vertical).

8.2 VIBRATION DIAGNOSIS

Spectrums (shown in Figures 8.7–8.10) are analyzed to find out for root cause for high vibration level of fan casing in horizontal and vertical directions. This database helped in the diagnosis of the bearing frequency analysis (mentioned in Section 8.3).

8.3 FREQUENCY ANALYSIS

There was increase in amplitude of fan casing offline vibration from 3.5 to 8.1 mm/s. Also, deviation in fan casing acceleration value from 3.2 to 6.0 g was observed (refer to Figure 8.11).

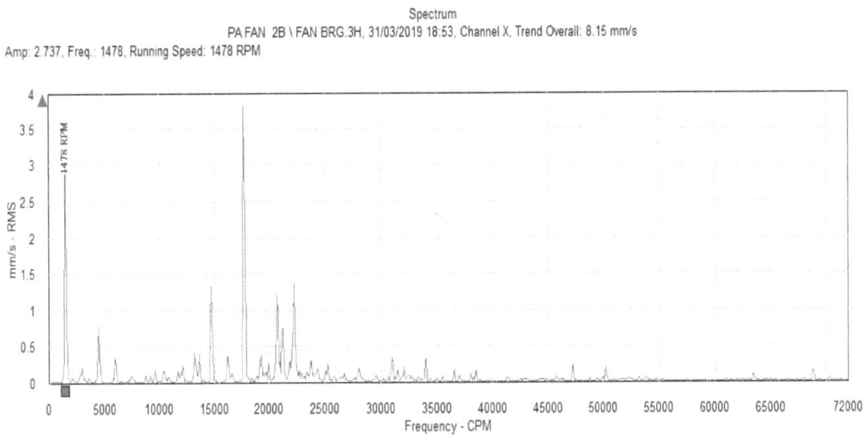

FIGURE 8.7 Spectrum of fan casing (horizontal).

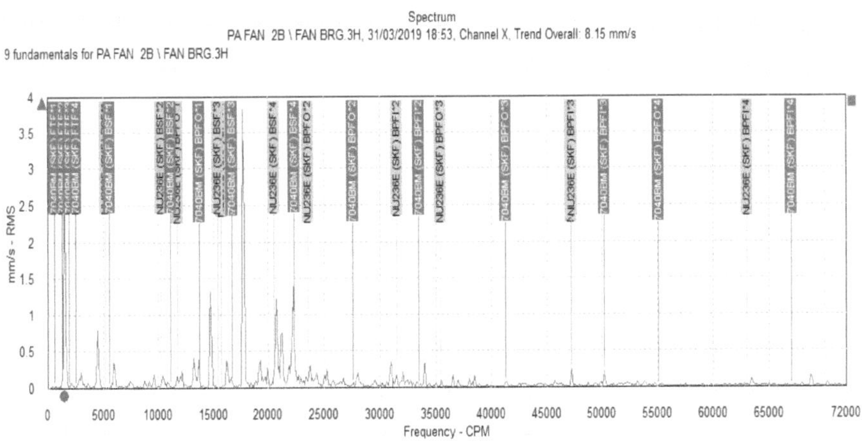

Spectrum Overall	8.157
Sync	7.009
Subsync	0.2224
Nonsync	4.167

FIGURE 8.8 Vibration spectrum of fan casing showing defect frequencies (at 22230 cpm).

With reference to Figures 8.7–8.11, spectrum analysis revealed an increase in predominant ball pass frequency inner race (BPFI) and ball spin frequency (BSF) of 17700 and 22230 cpm, respectively. There is an increase in higher frequencies and the contribution of non-synchronous energy (any vibration that is greater than the 1× frequency but is not a whole number multiple of it) is at higher side which is mainly due to bearing defects. The high frequencies got excited. These high frequencies were observed due to the bearing defect frequencies. Thus, it was confirmed that fan bearing deterioration is in progress.

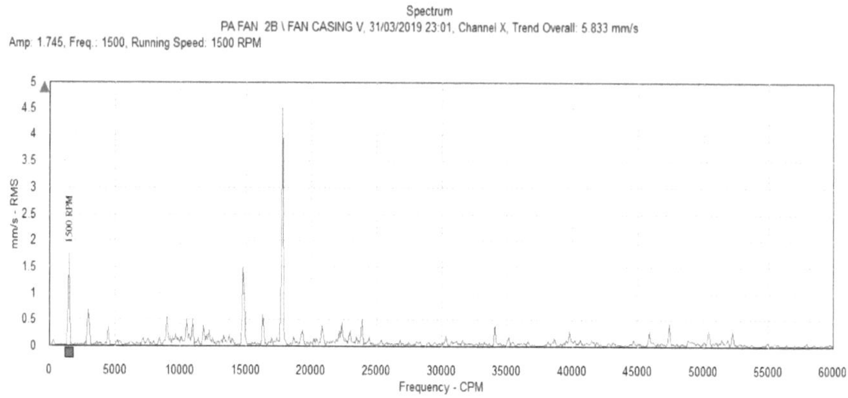

FIGURE 8.9 Spectrum of fan casing (vertical).

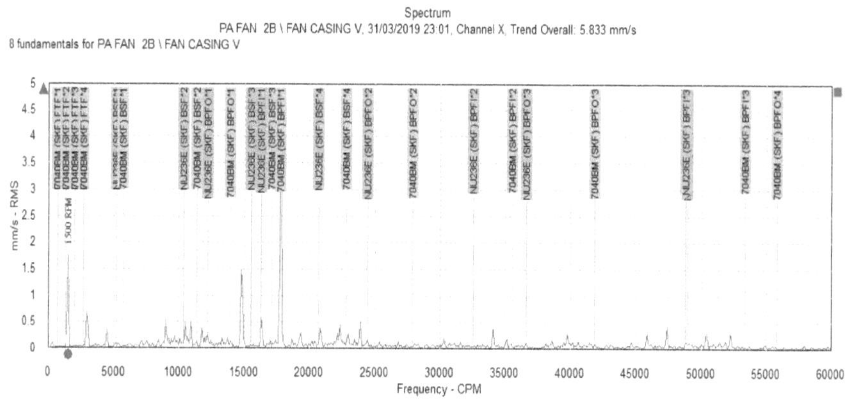

Spectrum Overall	5.833
Sync	2.174
Subsync	0.1309
Nonsync	5.412

FIGURE 8.10 Vibration spectrum of fan casing showing defect frequencies (17700 cpm).

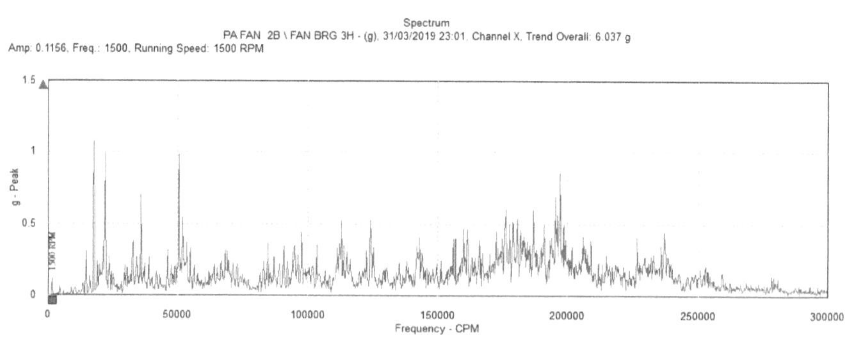

FIGURE 8.11 Acceleration spectrum of fan casing.

8.4 OBSERVATIONS DURING MAINTENANCE WORK

The condition of motor and fan side coupling was checked (see Figure 8.12). No deviation was noticed. Alignment between motor and fan was checked, and it was found good. Fan bearing clearances was found on higher side and scoring marks were observed on rollers as shown in Figure 8.13. Detailed inspection of all fan blades along with fan rotor and bearing housing was carried out. No deviation was observed (refer to Figure 8.14). Detailed inspection of fan shaft also carried out as shown in Figure 8.15. No abnormality was noticed.

FIGURE 8.12 Motor and fan side coupling inspection.

FIGURE 8.13 Damaged fan bearings.

FIGURE 8.14 Fan blades and rotor inspection.

FIGURE 8.15 Fan shaft inspection.

8.5 MAINTENANCE WORK PERFORMED

Motor was de-coupled from the rest of the group drive (see Figure 8.16). Before dismantling, initial alignment readings were taken for reference. All fan blades (46 nos.) were taken out and detailed inspection was carried out. Fan rotor also removed and replaced with a spare serviced one. All fan blades were installed on new fan rotor. Assembled rotor aligned with motor and coupling was done (see Figure 8.17). Trial was taken at full load, and all vibration data were found well within limit (Figures 8.18 and 8.19).

FIGURE 8.16 Replacement of fan rotor with spare serviced one.

FIGURE 8.17 Fan blade installation (46 nos.).

FIGURE 8.18 Final alignment work.

FIGURE 8.19 Fan ready after maintenance work.

8.6 POST-MAINTENANCE VIBRATION READINGS - OFFLINE

Table 8.1 shows the comparison of pre- and post-maintenance vibration readings which concludes the success of rotor replacement work.

8.7 POST-MAINTENANCE VIBRATION READINGS - ONLINE

Fan casing online vibration reduced from 11.2 to 2.9 mm/s (RMS) at full load condition as shown in Figure 8.20. This shows the comparison of pre- and post-maintenance online vibration readings which concludes the success of rotor replacement work.

TABLE 8.1
Vibration Report of PA Fan – 2B

			Initial Readings	After Replacing Fan Rotor
			31/03/2019	02/04/2019
Point		**Position**	Velocity (mm/s) RMS	
Motor	(NDE)	H	3.4	1.4
		V	0.7	0.5
		A	0.5	0.4
	Acceleration - g	H	0.4	0.3
Motor	(DE)	H	3.2	1.6
		V	0.5	0.4
		A	0.7	0.6
	Acceleration - g	H	0.6	0.3
Fan casing	(DE)	H	**8.1**	**1.6**
		V	**5.8**	**0.7**
		A	5.3	1.8
	Acceleration - g	H	**6.0**	**1.3**

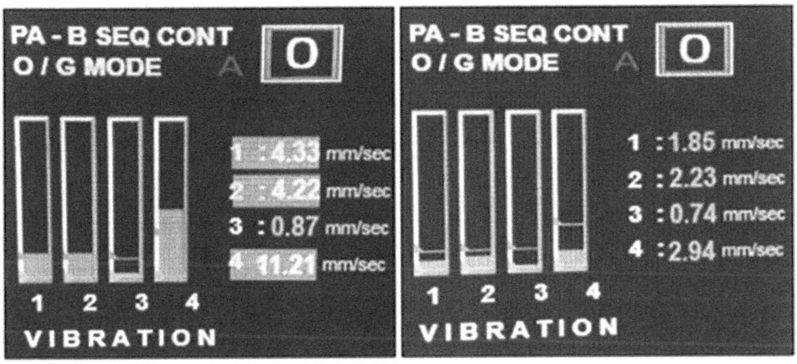

FIGURE 8.20 Pre- and post-maintenance online vibration data.

8.8 POST-MAINTENANCE VIBRATION SPECTRUM

Figures 8.21 and 8.22 show post-maintenance vibration spectrums for fan casing in horizontal and vertical directions which show no bearing defect peaks at 23300 and 17700 cpm respectively.

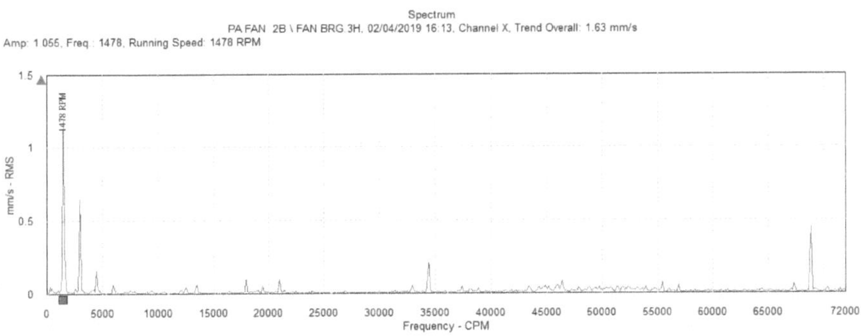

FIGURE 8.21 Spectrum of fan casing (horizontal).

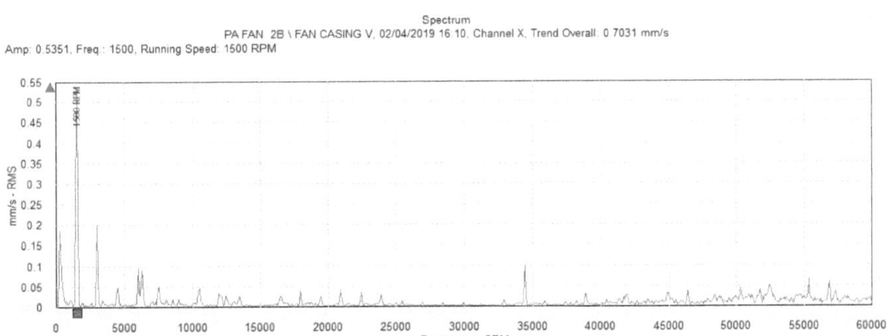

FIGURE 8.22 Spectrum of fan casing (vertical).

8.9 CONCLUSIONS

Primary Air (PA) fan is critical equipment that work on mainly two functions i.e. drying and transportation of coal to the furnace. Boiler is provided with two nos. of PA fans. Each PA fan air is provided with blade pitch control for controlling the loading on fans, outlet damper for isolation to facilitate start-up/maintenance of fan.

Since PA fan has no standby, the reliability of fan needs to be very high. It is a critical auxiliary in the process, failure of PA fan will directly affect generation loss of 0.09 MUs/h which costs revenue of around Rs.360,000 per hour. This made PA fan important equipment in the system. If any fan rotor fails during operation, it will lead to generation loss and in extreme case tripping of plant and interruption in power supply.

This case study has clearly demonstrated that condition-based vibration testing is a vital component in rotating machinery analysis and detection of abnormality at an early stage. Vibration monitoring and analysis helped in diagnosing the exact root cause of abnormity well in advance and avoided the secondary damage to fan and motor. Other major benefits include prevention of downtime, avoidance of major fan rotor repair or replacement costs and reliability of unit.

9 Case Study 8
Turbogenerator 01: In-Situ Balancing – January 2019

EXECUTIVE SUMMARY

The present case study is the success story of the implementation of condition monitoring techniques at ADTPS, where the in-situ balancing of Turbogenerator (TG) 01 is discussed. During annual overhaul period from 23 December 2018 to 24 January 2019, it was decided to replace high-pressure turbine (HPT) module with a new balanced one. On 20 January 2019, during initial rolling at 3000 rpm (before synchronization), it was found that vibration level of intermediate pressure turbine (IPT) front shaft (2*x*) observed on higher side of 220 μ (*p* –*p*). To observe the vibration stability of machine, on 20 January 2019, the unit was synchronized and found that the vibration level of IPT front shaft was increased from 220 to 227 μ (*p* –*p*) at 191 MW. After 4 hours of operation, the vibration level of IPT front shaft was further increased from 227 to 244 μ (*p* –*p*) and vibration level of high-pressure front 1*x* and 1*y* shaft deviated from 121 and 115 μ to 177 and 174 μ (*p* –*p*) respectively. No deviation was observed at all other bearing and shaft vibrations of low-pressure turbine (LPT), generator and exciter. It was decided to keep a close watch on all turbine process parameters. At 215 MW load condition, the vibration level of HP front 1*x* and 1*y* shaft was maintained on higher side of 180 μ (*p* –*p*) and the vibration level of IPT front shaft 2*x* was maintained at 280 μ (*p* –*p*). By observing the rising vibration trend of IPT front shaft 2*x* with an increase in load, unit withdrawn on 21 January 2019 for turbine in-situ balancing.

From 21 January 2019 to 24 January 2019, with the help of vibration and phase data analysis, 07 weights of a total of 770 g were added to IPT front at 150° and 05 weights of a total of 450 g were added to HPT front at 330° from zero marked position respectively.

After balancing, the vibration of IPT front shaft 2*x* was reduced from 280 to 149 μ (*p*–*p*) and HPT front shaft 1*x* and 1*y* were reduced from 168 to 139 and 73 to 64 μ (*p*–*p*). No other fluctuations were noticed in any vibration and process parameter reading.

9.1 PRE-MAINTENANCE HISTORY

During the overhaul period from 23 December 2018 to 24 January 2019, it was decided to replace the HP module with a new one. This new HP rotor was balanced on 9 May 2017 as per standard – ST 37008 – REV – 03 at BHEL, Hyderabad. Final weight with position on rotor is mentioned in Table 9.1 after tunnel balancing. Final balancing vibration details are shown in Figure 9.1. On 20 January 2019, during initial turbine rolling at 3000 rpm (before synchronization), it was found that IPT front shaft (2*x*) vibration level was on higher side at 220 μ (*p* –*p*) as depicted in Figure 9.2. Also, Figure 9.2 shows all other important process parameters of turbine and generator at the time of high vibration condition. To observe the vibration stability of machine, on 20 January 2019, the unit was synchronized and found that IPT front shaft 2*x* vibration level was increased from 220 to 227 μ (*p* –*p*) at 191 MW (see Figure 9.3). After

DOI: 10.1201/9781003432616-9

TABLE 9.1
Balancing Weights on Turbine Rotor

	Front End	Rear End
Weights of 40 g	6 Nos at 360°	9 Nos at 310°
	2 Nos at 75°	2 Nos at 75°
Total no. of weights	8 Nos	11 Nos

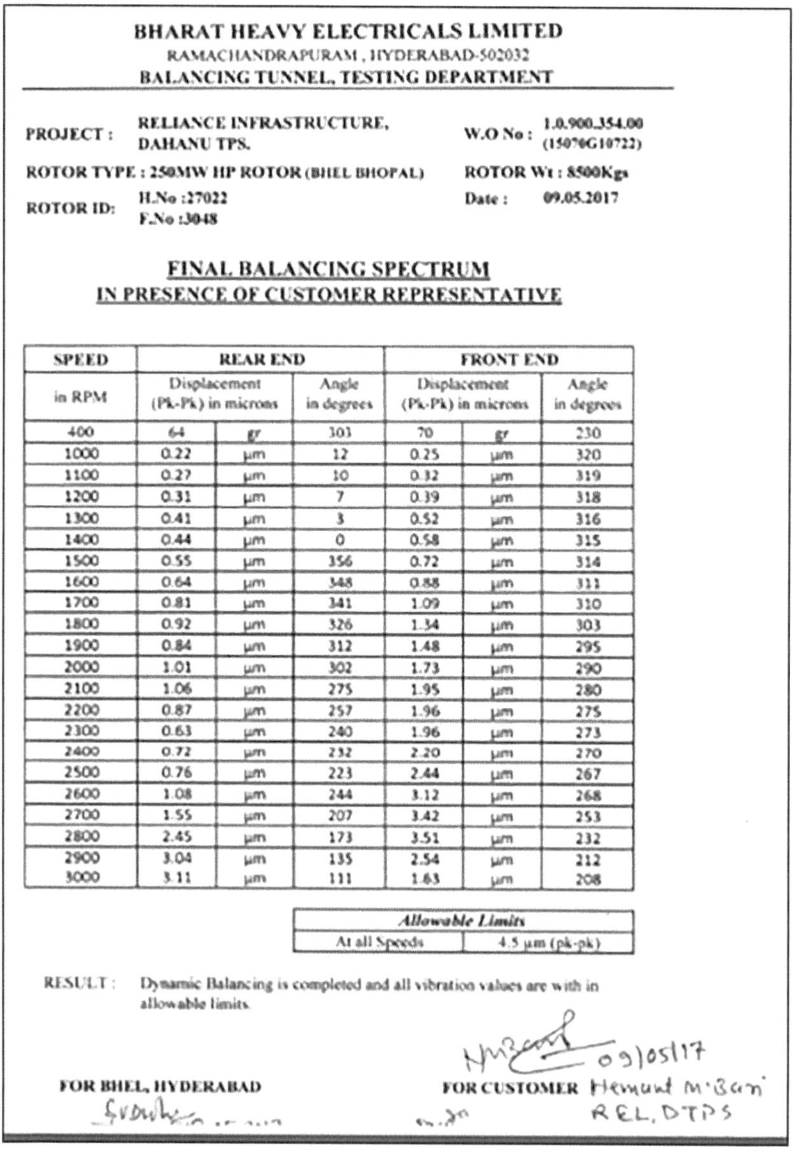

BHARAT HEAVY ELECTRICALS LIMITED
RAMACHANDRAPURAM , HYDERABAD-502032
BALANCING TUNNEL, TESTING DEPARTMENT

PROJECT : RELIANCE INFRASTRUCTURE, DAHANU TPS. **W.O No :** 1.0.900.354.00 (15070G10722)
ROTOR TYPE : 250MW HP ROTOR (BHEL BHOPAL) **ROTOR Wt :** 8500Kgs
ROTOR ID: H.No :27022 F.No :3048 **Date :** 09.05.2017

FINAL BALANCING SPECTRUM
IN PRESENCE OF CUSTOMER REPRESENTATIVE

SPEED	REAR END			FRONT END		
in RPM	Displacement (Pk-Pk) in microns		Angle in degrees	Displacement (Pk-Pk) in microns		Angle in degrees
400	64	gr	303	70	gr	230
1000	0.22	µm	12	0.25	µm	320
1100	0.27	µm	10	0.32	µm	319
1200	0.31	µm	7	0.39	µm	318
1300	0.41	µm	3	0.52	µm	316
1400	0.44	µm	0	0.58	µm	315
1500	0.55	µm	356	0.72	µm	314
1600	0.64	µm	348	0.88	µm	311
1700	0.81	µm	341	1.09	µm	310
1800	0.92	µm	326	1.34	µm	303
1900	0.84	µm	312	1.48	µm	295
2000	1.01	µm	302	1.73	µm	290
2100	1.06	µm	275	1.95	µm	280
2200	0.87	µm	257	1.96	µm	275
2300	0.63	µm	240	1.96	µm	273
2400	0.72	µm	232	2.20	µm	270
2500	0.76	µm	223	2.44	µm	267
2600	1.08	µm	244	3.12	µm	268
2700	1.55	µm	207	3.42	µm	253
2800	2.45	µm	173	3.51	µm	232
2900	3.04	µm	135	2.54	µm	212
3000	3.11	µm	111	1.63	µm	208

Allowable Limits	
At all Speeds	4.5 µm (pk-pk)

RESULT : Dynamic Balancing is completed and all vibration values are with in allowable limits.

09/05/17

FOR BHEL, HYDERABAD

FOR CUSTOMER Hemant M. Bari
REL, DTPS

FIGURE 9.1 Vibration readings after balancing.

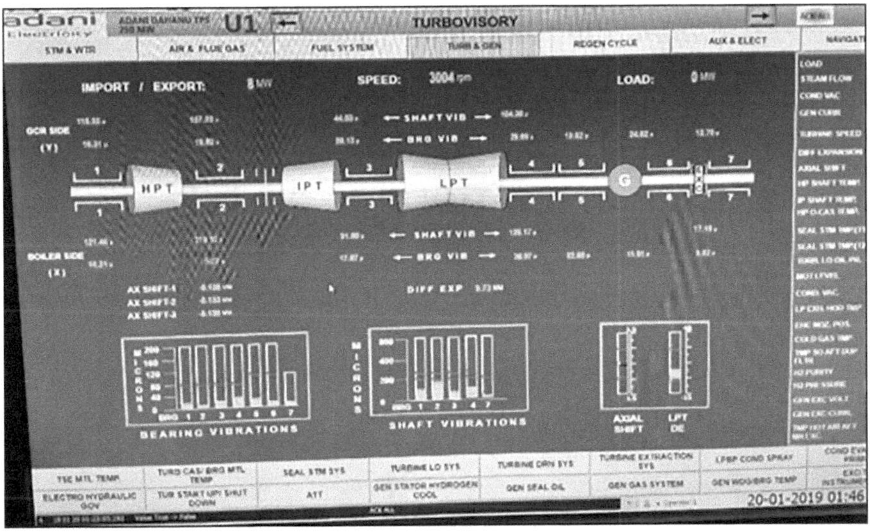

FIGURE 9.2 Turbine and generator (TG) vibration data on 20 January 2019 at 0 MW load.

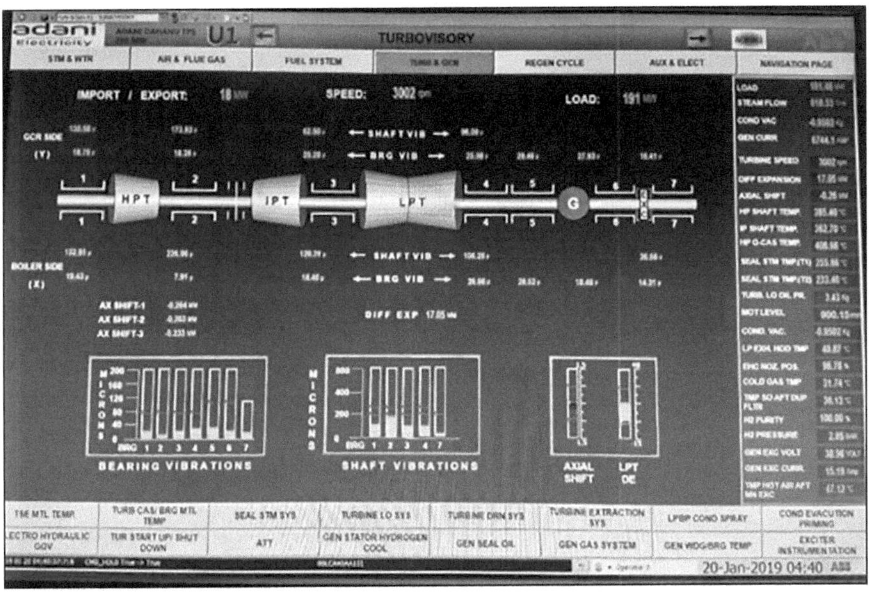

FIGURE 9.3 Turbine and generator (TG) vibration data on 20 January 2019 at 191 MW load.

4 hours of operation, IPT front shaft 2x vibration level further increased from 227 to 244 µ (p –p) and HP front 1x and 1y shaft vibration were deviated from 121 and 115 to 177 and 174 µ (p –p) respectively (see Figure 9.4). No deviations were observed in offline bearing casing vibrations (for all 06 bearings) at 190 MW load condition. As shown in Table 9.2, all casing offline vibration readings of turbine and generator were found within limits. Offline vibration spectrum analysis was carried out for having

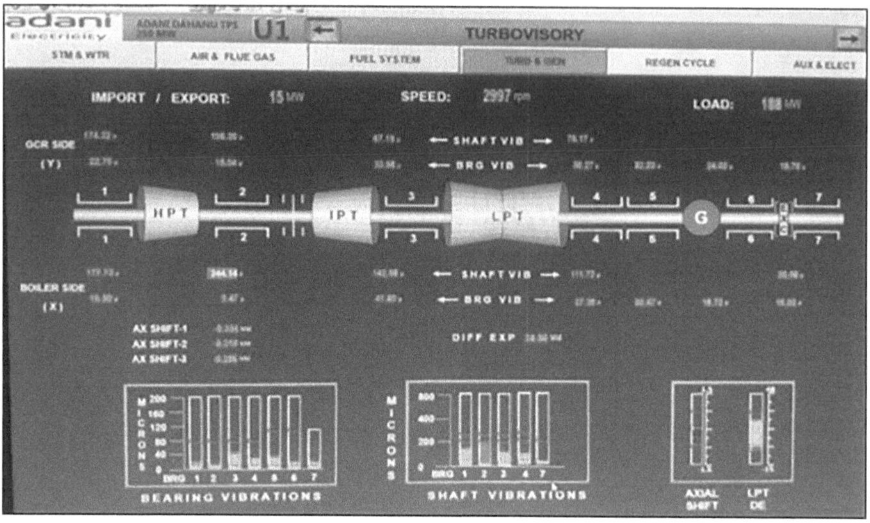

FIGURE 9.4 Turbine and generator (TG) vibration data at 190 MW load (after 04 h).

TABLE 9.2
Offline TG Bearing Casing Vibration Data

TG 01	Load – 190 MW	
Bearing No	**mm/s (RMS)**	
1	H	1.6
(HP turbine front)	V	0.7
	A	2.9
2	H	1.4
(IP turbine front)	V	1.3
	A	1.4
3	H	1.8
(IP turbine rear)	V	3.3
	A	5.7
4	H	2.0
(LP turbine rear)	V	4.9
	A	12.8
5	H	1.7
(Generator front)	V	5.3
	A	1.6
6	H	1.8
(Generator rear)	V	3.9
	A	1.7

high shaft vibration bearing 01 and 02 casing (Horizontal). No abnormal frequencies were noticed (Figure 9.5).

No deviation was observed at all other bearing and shaft vibrations of low-pressure turbine (LPT), generator and exciter. It was decided to keep a close watch on all

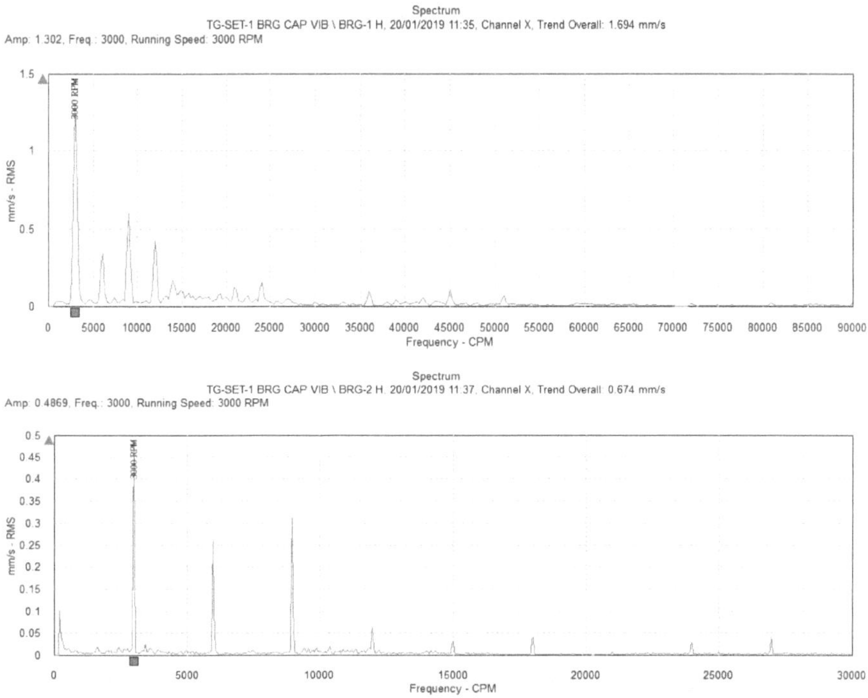

FIGURE 9.5 Vibration spectrum bearing 01 (LHS) and 02 (RHS) (horizontal).

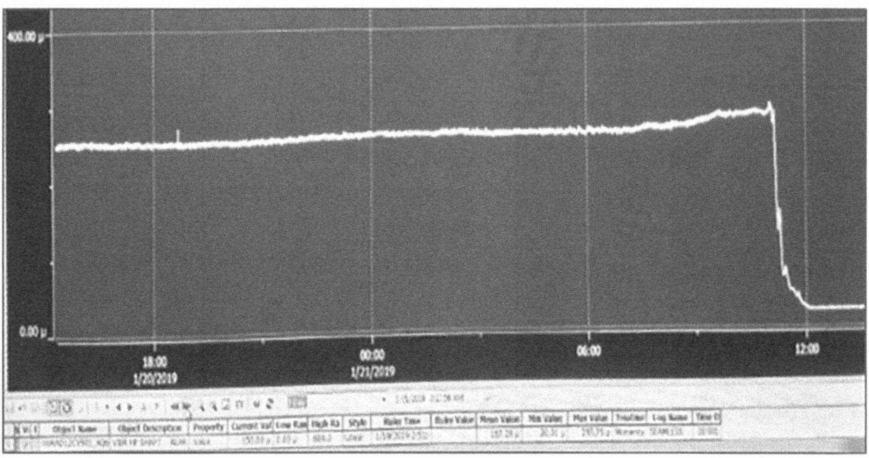

FIGURE 9.6 Turbine coast down vibration data.

turbine process parameters. At 215 MW, HP front 1*x* and 1*y* shaft vibrations were maintained on higher side of 180 µ (*p* –*p*) and IPT front shaft 2*x* vibration was maintaining 280 µ (*p* –*p*). By observing rising trend of IPT front shaft 2*x* vibration with increase in load, unit withdrawn on 21 January 2019 for turbine in-situ balancing. (Refer to Figure 9.6 showing turbine coast down vibration data).

A detailed report is enclosed herewith (Annexure 9.8 and 9.9).

9.2 PRE-BALANCING DATA COLLECTION SETUP

Initial setup was arranged for data collection such as installation of reflective tape at zero position on shaft area between bearing no. 04 and 05, setting laser tacho for speed monitoring, analyzer connection with JB cables and availability of data before rolling (Figure 9.7). On 22 January 2019, the turbine was rolled at 600 rpm and initial readings were taken as mentioned in Table 9.3. Data recorded for preliminary reference using offline vibration analyzer (Figure 9.8). On the same day, turbine was rolled at 3000 rpm and initial readings were taken as mentioned in Table 9.4. After analyzing the data mentioned in Table 9.4 and Figure 9.9, it was found that IP turbine front shaft $2x$ and $2y$ spectrum value of $1x$ component was high with phase difference of 103° which means close to 90° (as per standard) confirmed to go for single plane balancing.

9.3 PRE-BALANCING DATA COLLECTION

By analyzing the collected data (mentioned in Section 9.2) and considering the replacement of HP module during overhaul, it was decided to add high density

FIGURE 9.7 Pre-insitu balancing setup in progress.

TABLE 9.3
Vibration Data at 600 rpm Turbine Speed (Offline)

Bearing No.	Probe Location	Shaft Vibration (μm)	1 × Value (μm)	Phase Angle (Degrees)	Turbine Speed (RPM)
01	X	26.7	21.3	108	600
	Y	26.5	20.3	35	
02	X	61.4	45.9	63	
	Y	61.0	42.3	179	
03	X	35.6	22.3	154	
	Y	35.9	25.1	262	
04	X	56.0	17.9	244	
	Y	51.0	13.3	27	
Bearing 01 Offline casing vibration	H		0.103 mm/s (RMS)		
Bearing 02 Offline casing vibration	H		0.093 mm/s (RMS)		

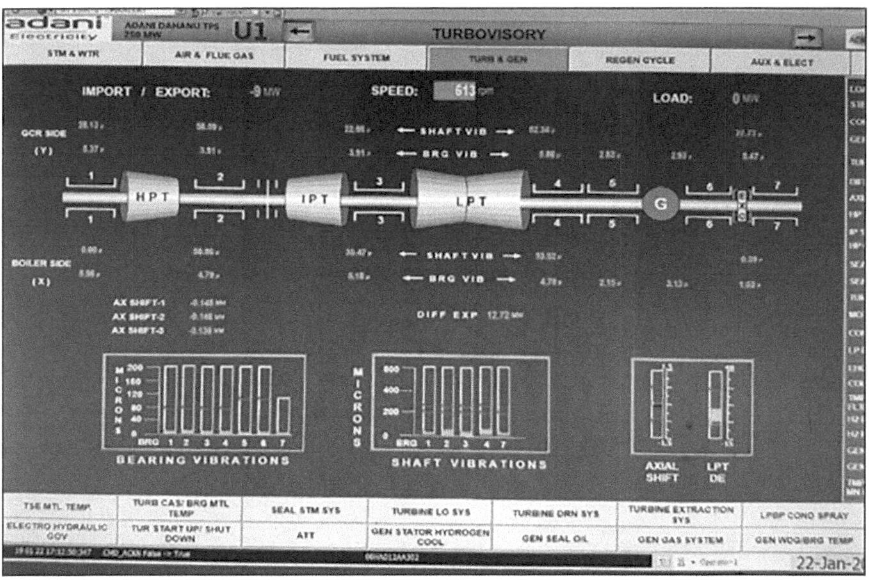

FIGURE 9.8 Vibration data at 600 rpm turbine speed (online).

TABLE 9.4
Vibration Data at 3000 rpm Turbine Speed (Offline)

Bearing No.	Probe Location	Shaft Vibration (μm)	1 × Value (μm)	Phase Angle (Degrees)	Turbine Speed (RPM)
01	X	79.3	65.6	227	3000
	Y	87.8	84.6	95	
02	X	225	**213**	**110**	
	Y	160	**141**	**213**	
03	X	98.4	89.7	219	
	Y	64.9	56.1	350	
04	X	122	77.3	42	
	Y	73	3.56	181	
Bearing 01 Offline casing vibration	H		**2.02 mm/s (RMS)**		
Bearing 02 Offline casing vibration	H		**0.57 mm/s (RMS)**		

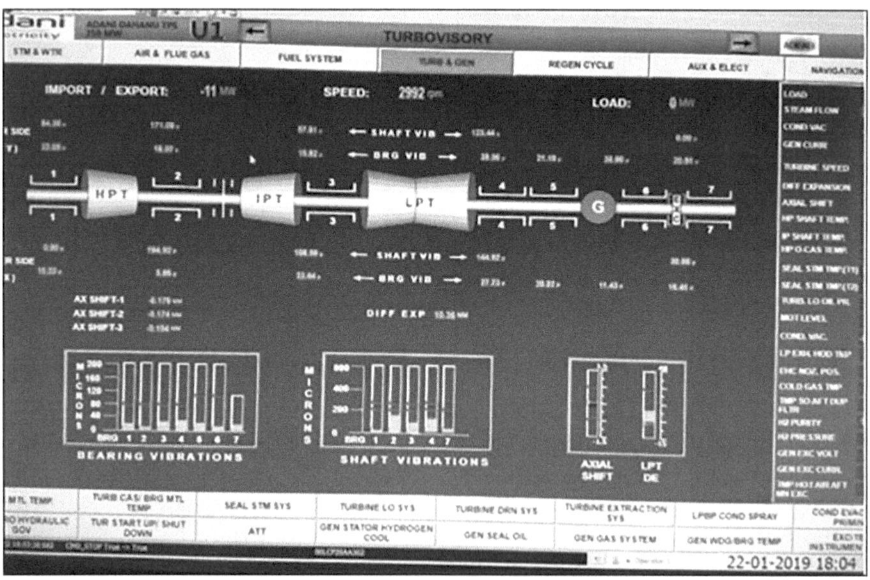

FIGURE 9.9 Vibration data at 3000 rpm turbine speed (online).

560 g (5 weights of 112 g) of weight at 150°on IP turbine front plane (5'oclock position) in the opposite direction of rotation from 0° marking on the shaft. Accordingly, the desired position was set, and marking was carried out as shown in Figure 9.10. Also, weights were added to the intermediate pressure turbine front side.

On 22 January 2019, turbine was rolled at 3000 rpm and again data were collected (as presented in Table 9.5 and Figure 9.11) for offline and online respectively.

FIGURE 9.10 Balancing weights addition to IPT front side.

TABLE 9.5
Vibration Data after Adding Weights at IPT Front (Offline)

Bearing No.	Probe Location	Shaft Vibration (μm)	1 × Value (μm)	Phase Angle (Degrees)	Turbine Speed (RPM)
01	X	68.3	55.2	223	3000
	Y	68.2	63.7	87	
02	X	155	139	105	
	Y	134	112	215	
03	X	67.3	55.9	224	
	Y	55.7	45.6	343	
04	X	115	68.4	20	
	Y	85.4	40.2	136	
Bearing 01 Offline casing vibration	H		**1.0 mm/s (RMS)**		
Bearing 02 Offline casing vibration	H		**1.1 mm/s (RMS)**		

9.4 CORRECTION RUN - BY ADDING 224 G AT 150° IP TURBINE (FRONT)

By observing a positive response from rotor, reduction in IP turbine front shaft 2x vibration from 225 to 155 µ(p –p). It was decided to add 02 more weights of 112 g each at the same location. After adding weights, on 23 January 2019, turbine was again rolled at 3000 rpm and data were collected and presented in Table 9.6.

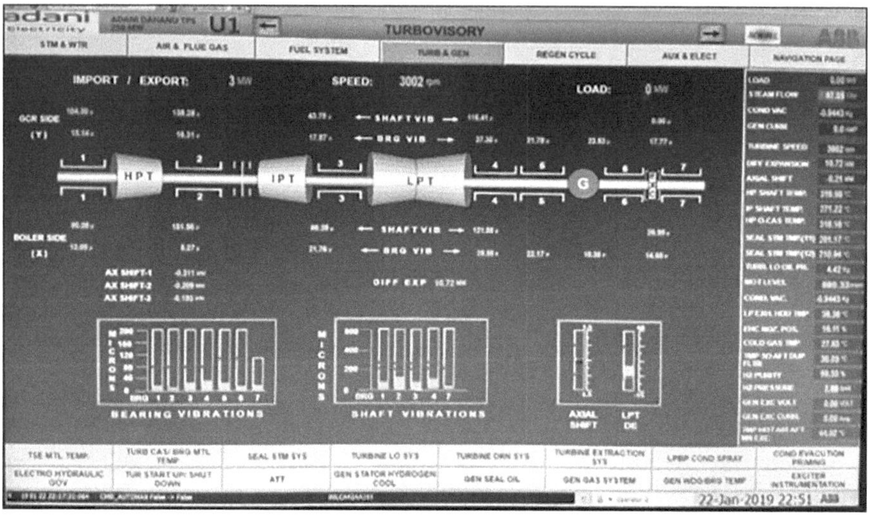

FIGURE 9.11 Vibration data after adding weights at IP turbine front (offline).

TABLE 9.6
Vibration Data after Adding Additional Weights at IP Turbine Front (Offline)

Bearing No.	Probe Location	Shaft Vibration (µm)	1 × Value (µm)	Phase Angle (Degrees)	Turbine Speed (RPM)
01	X	73.5	65.5	280	3000
	Y	93.3	89.6	115	
02	X	151	115	103	
	Y	120	97.2	216	
03	X	71.9	58.8	235	
	Y	53.2	42.2	341	
04	X	113	71.7	19	
	Y	83.7	46.3	136	
Bearing 01 Offline casing vibration	**H**		**0.99 mm/s (RMS)**		
Bearing 02 Offline casing vibration	**H**		**0.74 mm/s (RMS)**		

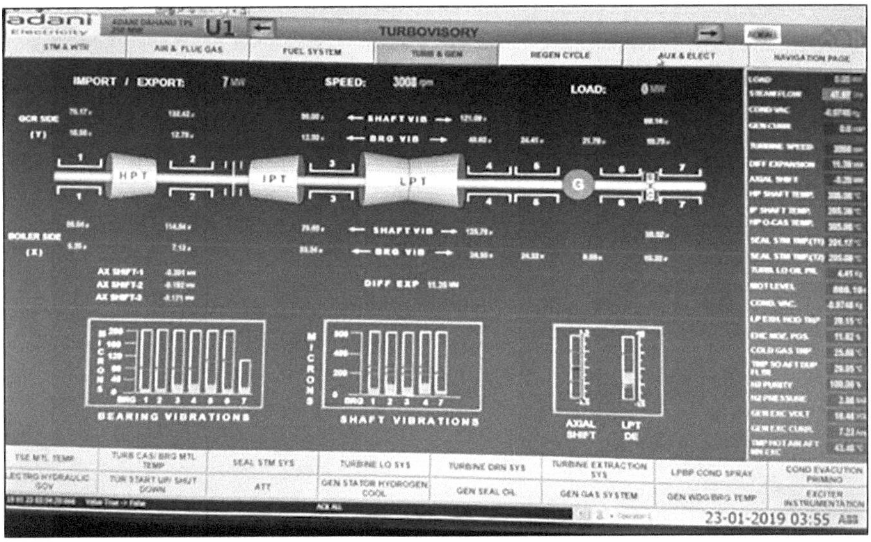

FIGURE 9.12 Vibration data after adding additional weights at IP turbine front (online).

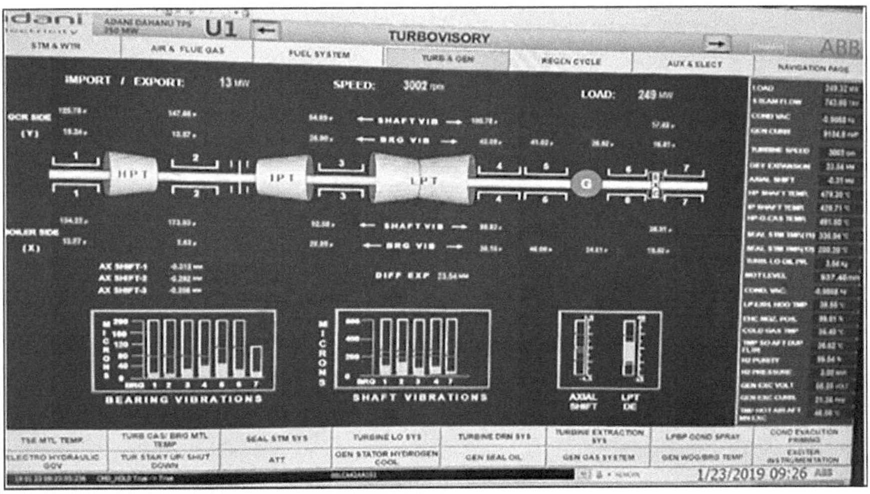

FIGURE 9.13 Vibration data at 250 MW load after adding weights at IP turbine front.

By observing further reduction in all shaft vibrations, it was decided to synchronize the machine. On 23 January 2019, the unit was synchronized, and the vibration level monitored. IP turbine front shaft 2*x* vibration was found as 173 μ (*p* –*p*) at 250 MW load (refer to Figure 9.13).

With reference to Table 9.7 and Figure 9.14, on 23 January 2019, IP turbine front shaft 2*x* vibration was found as 167 μ (*p* –*p*) at 250 MW load (after 03 hours stability).

TABLE 9.7
Vibration Data at 250 MW Load after Adding Weights at IP Turbine Front (Offline)

Bearing No.	Probe Location	Shaft Vibration (μm)	1 × Value (μm)	Phase Angle (Degrees)	Load (MW)
01	X	135	115	284	250
	Y	126	106	107	
02	X	174	147	105	
	Y	152	125	206	
03	X	91.1	79.9	238	
	Y	70.5	56.2	335	
04	X	105	95.3	334	
	Y	79.8	53.2	101	
Bearing 01 Offline casing vibration	H		**1.46 mm/s (RMS)**		
Bearing 02 Offline casing vibration	H		**1.27 mm/s (RMS)**		

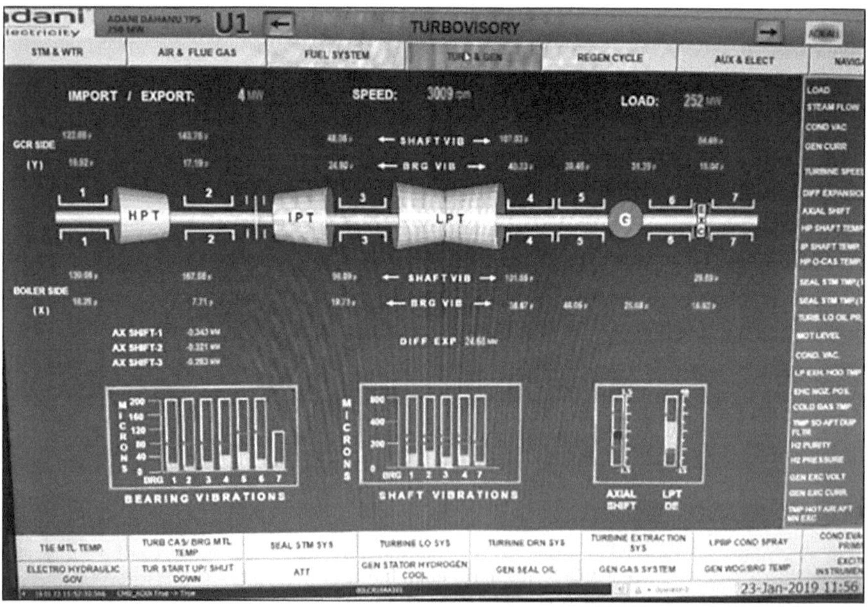

FIGURE 9.14 Vibration data at 250 MW load after adding weights at IP turbine front (online).

After 07 hours of operation, on 23 January 2019, it was noticed that IP turbine front shaft $2x$ vibration was found as 169 μ (p –p) and HP turbine front shaft $1y$ vibration was increased from 122 to 168 μ (p –p) at 190 MW load.

9.5 FINAL CORRECTION RUN - BY ADDING 450 G AT 150° HP TURBINE (FRONT)

By seeing deviations in HP front shaft $1y$ vibration at 190 MW from 122 to 168 μ (p-p), it was decided to go for one more correction. Unit withdrawn on 24 January 2019 for final correction. After analyzing the phase data collected at 190 MW, it was found that HP turbine and IP turbine front shaft $1x$ and $2x$ spectrum value of $1x$ component was high with a phase difference of 183° which confirmed to go for two plane balancing (see Table 9.8 and Figure 9.15).

With reference to discussion held in HoD meeting, it was finalized to add high density 450 g (5 weights of 90 g) of weight at 330° on HP turbine front plane (11 o'clock position) in the opposite direction of rotation from 0° marking on shaft. Weights were added to HP turbine front side (Figures 9.16–9.18).

After adding weights at HP turbine front side, on 24 January 2019, turbine was rolled to 3000 rpm and data collected and presented in Table 9.9.

9.6 LOAD READINGS – (AFTER BALANCING)

After balancing, on 24 January 2019, vibration readings were taken on full load (250 MW). Presented in Table 9.10 and Figure 9.19.

TABLE 9.8
Vibration Data at 190 MW Load after Adding Weights at IP Turbine Front (Offline)

Bearing No.	Probe Location	Shaft Vibration (μm)	1× Value (μm)	Phase Angle (Degrees)	Load (MW)
01	X	139	**127**	**296**	190
	Y	164	146	117	
02	X	181	**142**	**110**	
	Y	142	105	218	
Bearing 01 Offline casing vibration	**H**		**1.0 mm/s (RMS)**		
Bearing 02 Offline casing vibration	**H**		**1.1 mm/s (RMS)**		

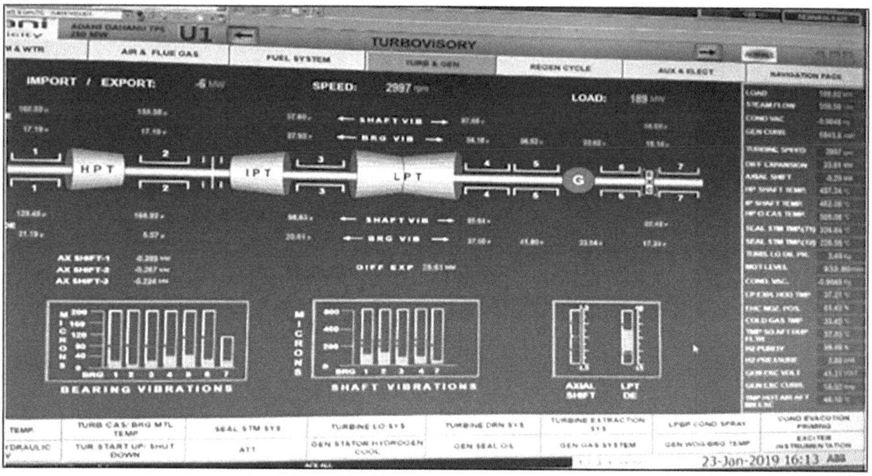

FIGURE 9.15 Vibration data at 190 MW load after adding weights at IP turbine front (online).

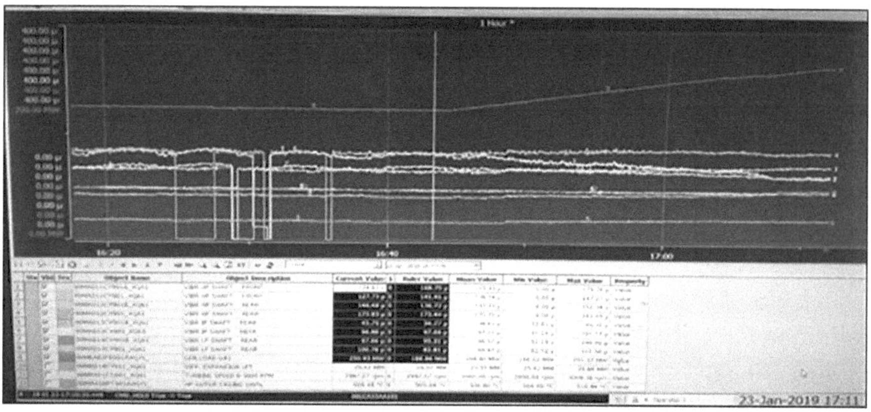

FIGURE 9.16 Online shaft vibration trend.

FIGURE 9.17 Weights addition at HP turbine front side.

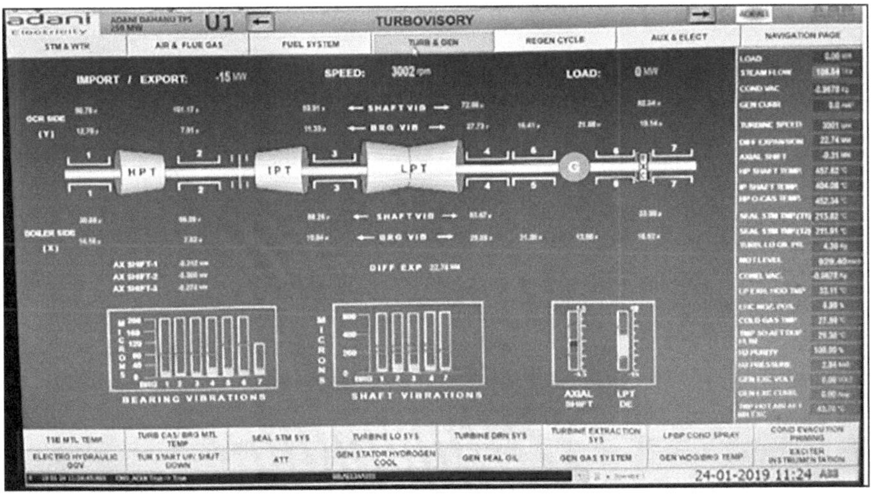

FIGURE 9.18 Vibration data after adding weights at HP turbine front (online).

TABLE 9.9
Vibration Data after Adding Weights at HP Turbine Front (Offline)

Bearing No.	Probe Location	Shaft Vibration (μm)	1× Value (μm)	Phase Angle (Degrees)	Turbine Speed (RPM)
01	X	22.0	8.31	339	3000
	Y	50.4	43.9	137	
02	X	97.5	84.2	103	
	Y	101	80.1	200	
03	X	86.8	76.3	223	
	Y	62.6	48.4	322	
04	X	67.1	41.0	358	
	Y	54.1	25.2	98	
Bearing 01 Offline casing vibration	**H**		**0.91 mm/s (RMS)**		
Bearing 02 Offline casing vibration	**H**		**0.22 mm/s (RMS)**		

9.7 RESULTS AND DISCUSSION

All shaft and bearing vibrations were remained stable for the next 24 hours. After balancing, IP turbine front shaft vibration 2x was reduced from 280 to 149 μ (p–p). Also, HP turbine front shaft vibrations 1x and 1y were reduced from

TABLE 9.10
Vibration Data at 250 MW Load (after Balancing - Offline)

Bearing No.	Probe Location	Shaft Vibration (μm)	1× Value (μm)	Phase Angle (Degrees)	Load (MW)
01	X	70.3	48.2	293	250
	Y	80.1	65.1	123	
02	X	156	127	102	
	Y	145	114	201	
03	X	85.1	69.1	239	
	Y	63.7	51.7	330	
04	X	89.2	75.3	348	
	Y	73.2	48.3	102	
Bearing 01 Offline casing vibration	H		0.99 mm/s (RMS)		
Bearing 02 Offline casing vibration	H		1.11 mm/s (RMS)		

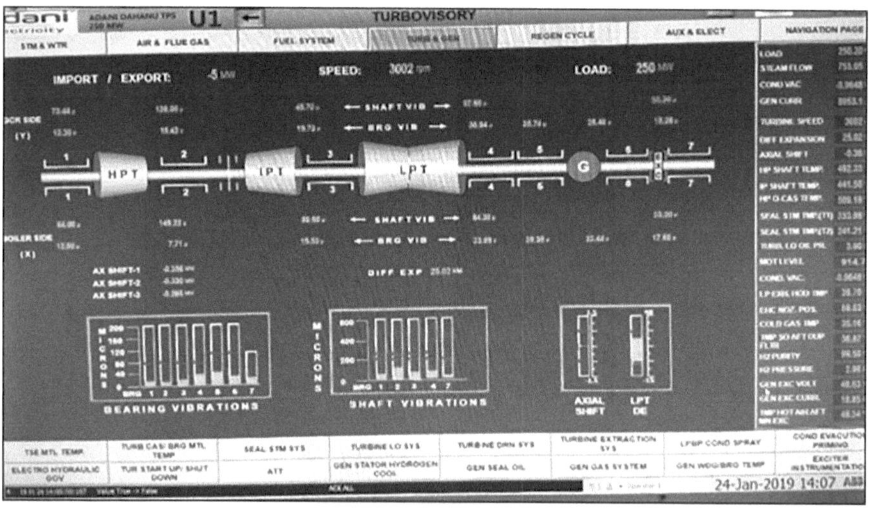

FIGURE 9.19 Vibration data at 250 MW load (after balancing – online).

139 and 168 to 64 and 73 μ (*p*–*p*). No further fluctuations were noticed. Vibration readings of all bearing offline casing were found within the prescribed limit (Table 9.11).

Offline vibration spectrum analysis was carried out and no abnormal frequencies were observed as depicted in Figure 9.20. Observing stability in HP turbine and IP turbine front shaft and bearing vibrations, it was decided to stop in-situ balancing and concluded with no further corrections.

TABLE 9.11
Offline TG Bearing Casing Vibration (after Balancing – Offline)

TG 01	Load – 250 MW	
Bearing No	**mm/s (RMS)**	
1	H	0.9
(HP turbine front)	V	0.5
	A	2.6
2	H	1.1
(IP turbine front)	V	1.2
	A	0.3
3	H	0.9
(IP turbine rear)	V	2.0
	A	4.0
4	H	1.8
(LP turbine rear)	V	4.2
	A	4.1
5	H	2.1
(Generator front)	V	5.0
	A	1.9
6	H	1.5
(Generator rear)	V	3.2
	A	1.5

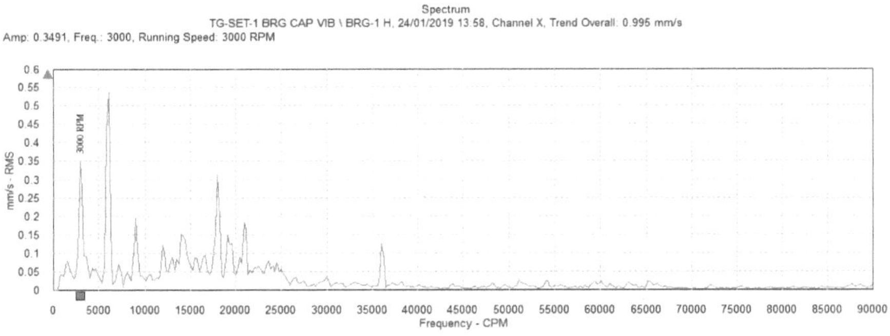

FIGURE 9.20 TG bearing 01 vibration spectrum (H)- (after balancing – offline).

9.8 ANNEXURE A - FINAL BALANCING PROTOCOL

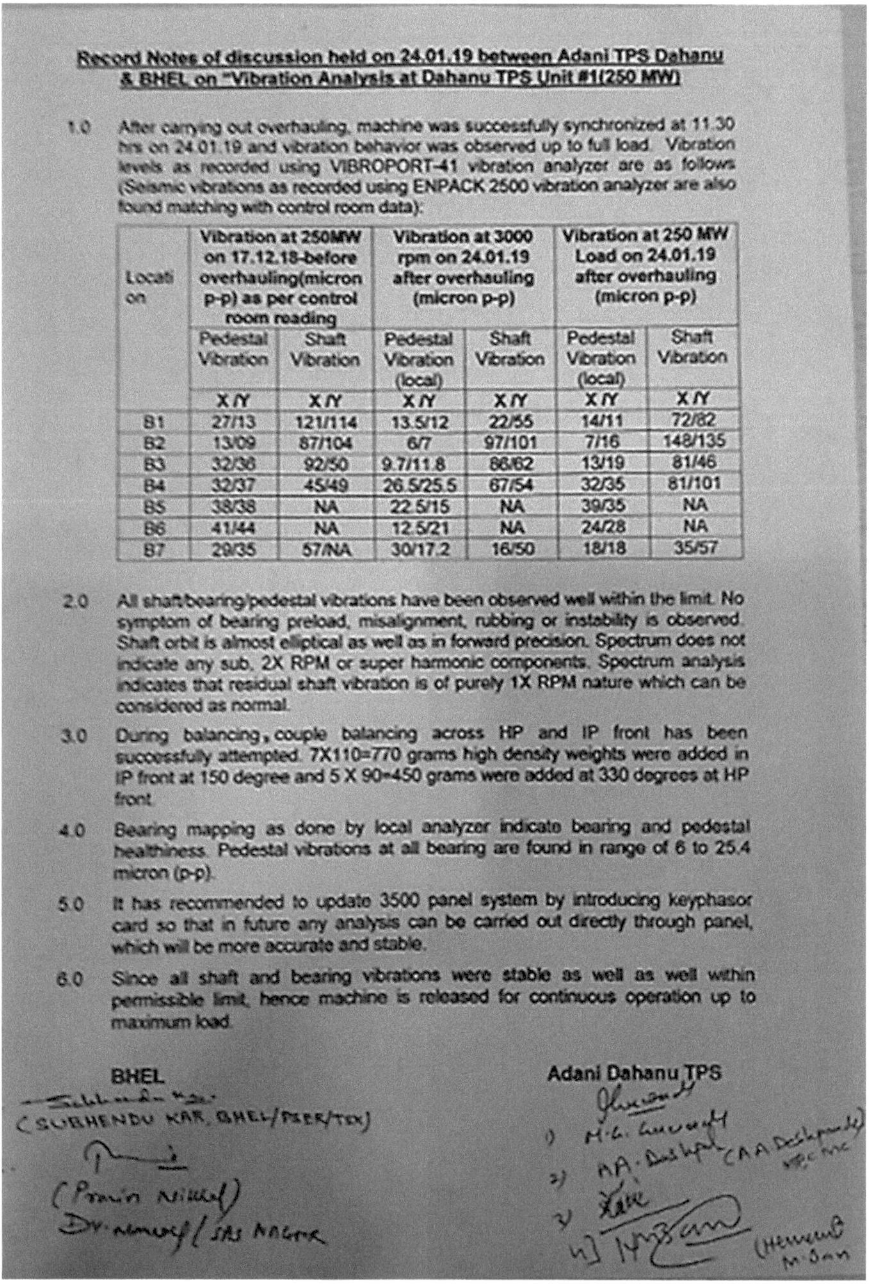

Record Notes of discussion held on 24.01.19 between Adani TPS Dahanu & BHEL on "Vibration Analysis at Dahanu TPS Unit #1(250 MW)

1.0 After carrying out overhauling, machine was successfully synchronized at 11.30 hrs on 24.01.19 and vibration behavior was observed up to full load. Vibration levels as recorded using VIBROPORT-41 vibration analyzer are as follows (Seismic vibrations as recorded using ENPACK 2500 vibration analyzer are also found matching with control room data):

Location	Vibration at 250MW on 17.12.18-before overhauling(micron p-p) as per control room reading		Vibration at 3000 rpm on 24.01.19 after overhauling (micron p-p)		Vibration at 250 MW Load on 24.01.19 after overhauling (micron p-p)	
	Pedestal Vibration	Shaft Vibration	Pedestal Vibration (local)	Shaft Vibration	Pedestal Vibration (local)	Shaft Vibration
	X /Y	X /Y	X /Y	X /Y	X /Y	X /Y
B1	27/13	121/114	13.5/12	22/55	14/11	72/82
B2	13/09	87/104	6/7	97/101	7/16	148/135
B3	32/36	92/50	9.7/11.8	86/62	13/19	81/46
B4	32/37	45/49	26.5/25.5	67/54	32/35	81/101
B5	38/38	NA	22.5/15	NA	39/35	NA
B6	41/44	NA	12.5/21	NA	24/28	NA
B7	29/35	57/NA	30/17.2	16/50	18/18	35/57

2.0 All shaft/bearing/pedestal vibrations have been observed well within the limit. No symptom of bearing preload, misalignment, rubbing or instability is observed. Shaft orbit is almost elliptical as well as in forward precision. Spectrum does not indicate any sub, 2X RPM or super harmonic components. Spectrum analysis indicates that residual shaft vibration is of purely 1X RPM nature which can be considered as normal.

3.0 During balancing, couple balancing across HP and IP front has been successfully attempted. 7X110=770 grams high density weights were added in IP front at 150 degree and 5 X 90=450 grams were added at 330 degrees at HP front.

4.0 Bearing mapping as done by local analyzer indicate bearing and pedestal healthiness. Pedestal vibrations at all bearing are found in range of 6 to 25.4 micron (p-p).

5.0 It has recommended to update 3500 panel system by introducing keyphasor card so that in future any analysis can be carried out directly through panel, which will be more accurate and stable.

6.0 Since all shaft and bearing vibrations were stable as well as well within permissible limit, hence machine is released for continuous operation up to maximum load.

BHEL

(SUBHENDU KAR, BHEL/PEER/TEX)

Adani Dahanu TPS

FIGURE 9.21 Annexure A.

9.9 ANNEXURE B - FINAL BALANCING WEIGHT CHART

07 weights of total 770 grams at 150º IPT (front) - (5'oclock position)

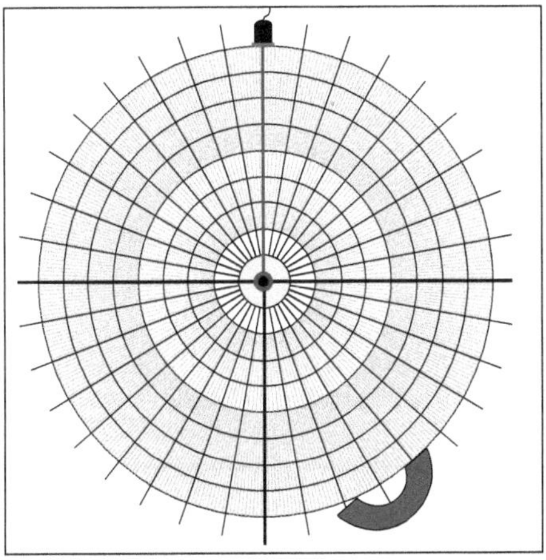

05 weights of total 450 grams at 330º HPT (front) - (11'oclock position)

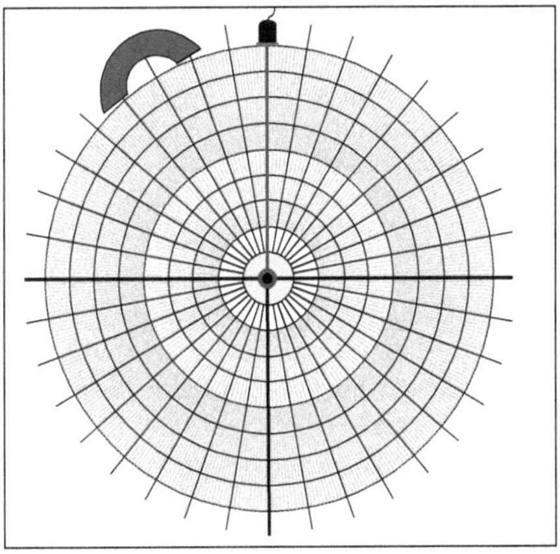

FIGURE 9.22 Annexure B.

10 Case Study 9 *SAC Compressor 2B: Motor Bearings Deterioration – July 2018*

EXECUTIVE SUMMARY

The present case study is the success story of the implementation of condition monitoring techniques at ADTPS, wherein imminent motor bearing deterioration of HT auxiliary Service Air Compressor – 2B was diagnosed.

This case study has clearly demonstrated that condition-based vibration monitoring is a vital component in rotating machinery analysis and detection of abnormality at an early stage. In the present case study, vibration spectrum analysis played a major supportive role in pinpointing the motor bearing deterioration.

Vibration monitoring and spectrum analysis helped in diagnosing the exact root cause of abnormity well in advance. This has saved the plant from huge losses in terms of generation, unplanned breakdown and associated maintenance costs.

On 29 June 2018, operator of compressor control room noticed deviation in vibration of motor drive end (DE) bearing from 4.4 to 5.2 mm/s (RMS) during the routine monitoring. After collecting vibration data, it was observed that horizontal and axial vibrations of motor NDE and DE bearing deviated from 3.6 to 4.4 and 2.2 to 3.4 mm/s (RMS) respectively. No other deviations were noticed in any process parameters and compressor vibrations. Spectrum analysis revealed that there was an increment in bearing outer race defect frequency and was found 7740 cpm. These high frequencies lead to the bearing defect frequencies. Thus, it was confirmed that motor bearing deterioration was in progress. It has been decided to keep under close monitoring with a limit value of up to 6.0 mm/s.

On 3 July 2018, vibration in motor DE bearing was increased to 5.8 mm/s. With such observation, it was decided to take shutdown of compressor in a planned way to replace motor bearings. During the maintenance, minor scoring marks were observed on motor NDE and DE bearings and the same were replaced. During the maintenance work, thorough inspection of motor rotor and stator winding were carried out and no damage/scoring marks were observed. After replacing both bearings, motor was aligned with compressor and a trial was taken at full load. Vibration in motor DE bearing was reduced from 5.8 to 0.8 mm/s (RMS).

10.1 PRE-MAINTENANCE HISTORY

On 29 June 2018, during routine monitoring, the operator of compressor control room noticed deviation in online vibration readings of motor DE bearing (horizontal) from 4.4 to 5.2 mm/s (as shown in Figure 10.1). Offline vibration data were collected, and it was observed that horizontal and axial vibration of motor non-drive end (NDE) and drive end (DE) bearing were deviated from 3.6 to 4.4 and 2.2 to 3.4 mm/s (RMS) respectively (see Figure 10.2). However, compressor vibrations were found normal,

DOI: 10.1201/9781003432616-10

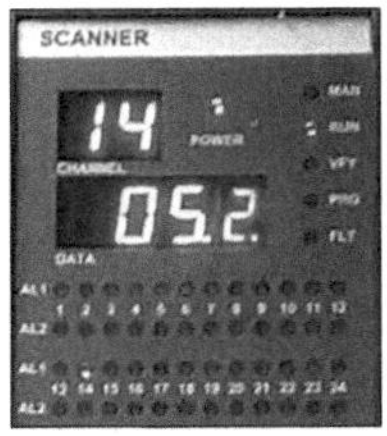

FIGURE 10.1 Motor DE – H online vibration reading on 29 June 2018.

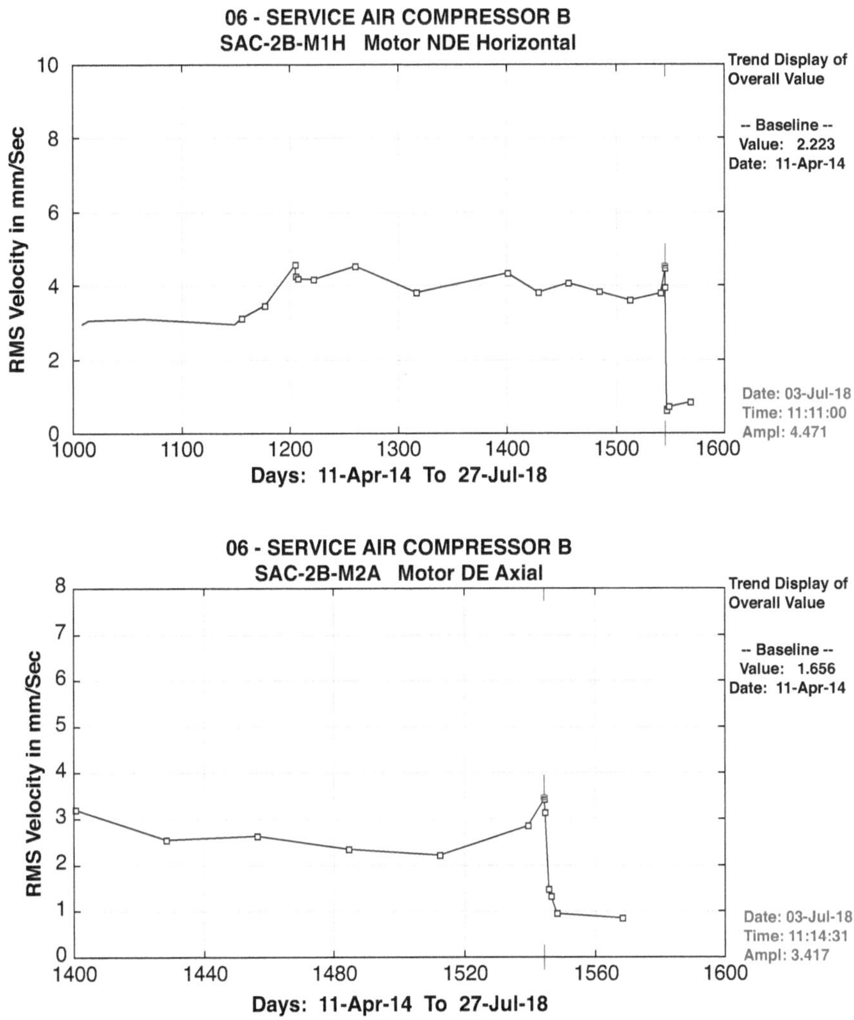

FIGURE 10.2 Vibration trend of Motor NDE (H) – LHS and Motor DE (A) – RHS.

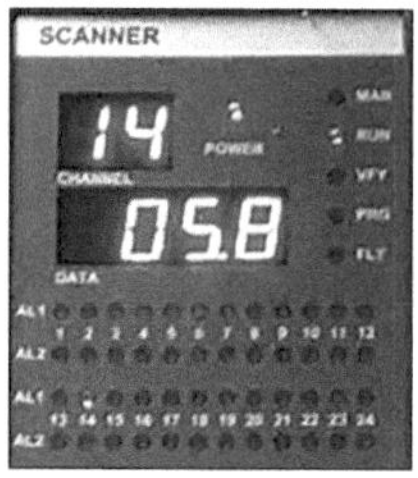

FIGURE 10.3 Motor DE – H online vibration reading on 03 July 2018.

and no deviations were noticed in any process parameters. Such high vibration issue of motor was discussed in HOD meeting. It was decided to keep under close monitoring with a limit value of up to 6.0 mm/s. On 3 July 2018, online vibration readings of motor DE bearing was found 5.8 mm/s (see Figure 10.3). Observing the increasing trend of motor vibration, it was decided to take planned outage of compressor for replacing motor bearings and work order (300000093153) was raised on SAP software to electrical maintenance department.

10.2 VIBRATION DIAGNOSIS

The spectrum analysis for root cause was investigated as follows:

With reference to Figures 10.4–10.7, spectrum analysis revealed that increase in predominant peak for ball pass frequency outer race (BPFO) bearing defect was found at 7740 cpm. There was increase in higher frequencies and the contribution of non-synchronous energy (any vibration that is greater than the 1× frequency but is not a whole number multiple of it) was found at higher side which was mainly due to bearing defects. The high frequencies got excited. These high frequencies were

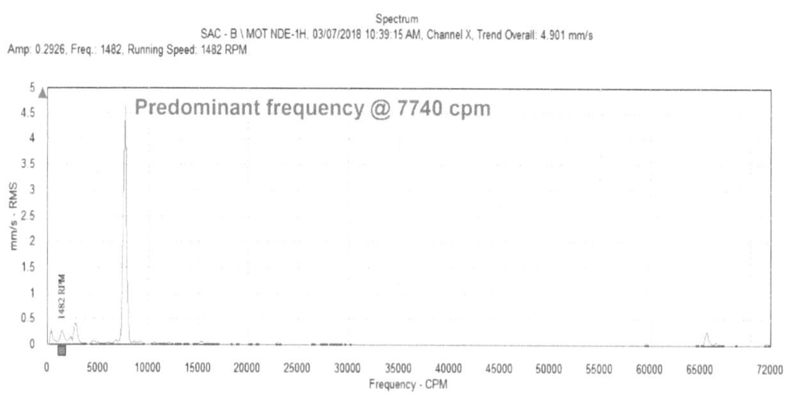

FIGURE 10.4 Vibration spectrum of motor NDE bearing - (horizontal).

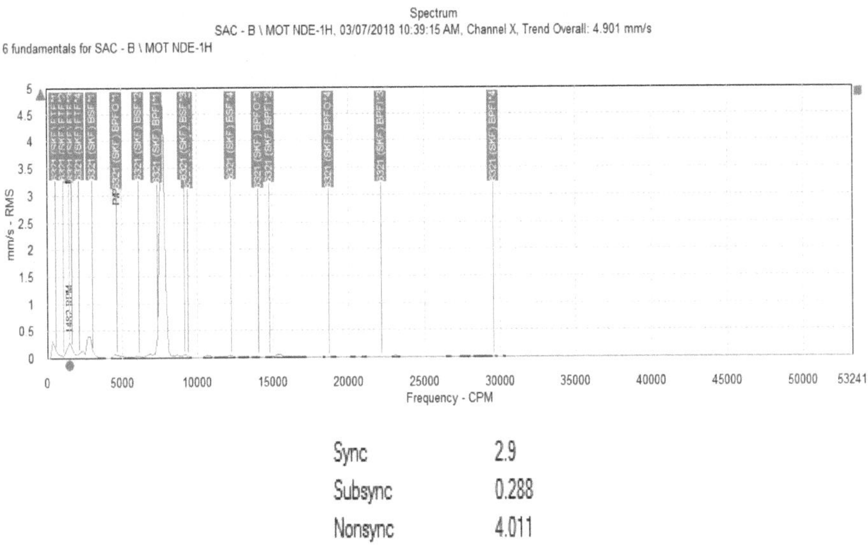

FIGURE 10.5 Vibration spectrum of NDE motor bearing showing defect frequencies.

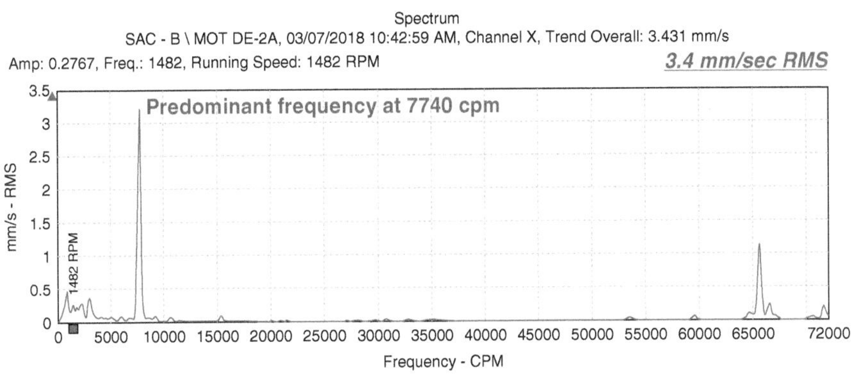

FIGURE 10.6 Spectrum of motor DE bearing (axial).

generated due to bearing defect. Thus, it was confirmed that motor bearing deterioration was in progress. It was concluded that vibration of motor NDE and DE bearing increased due to early-stage deterioration of bearings. Therefore, it was decided to stop compressor for replacement of both motor bearings.

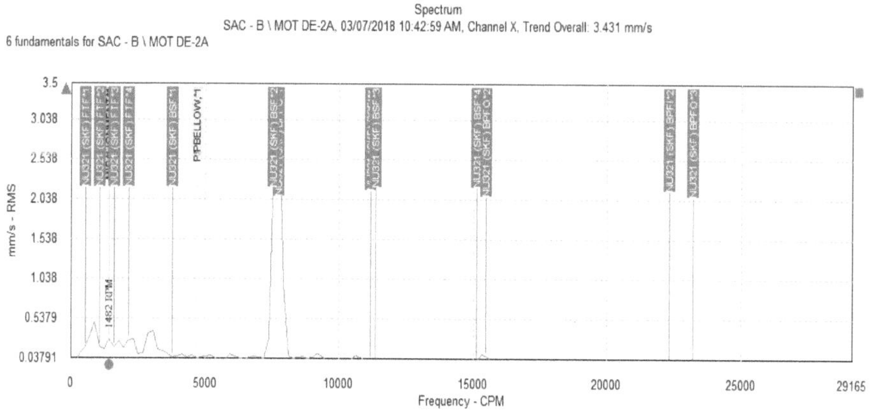

Sync	2.097
Subsync	0.5344
Nonsync	3.051

FIGURE 10.7 Vibration Spectrum of Motor DE bearing showing defect frequencies.

10.3 OBSERVATIONS RECORDED DURING MAINTENANCE WORK

Condition of motor pulley and belts was checked. No deviation was noticed. Alignment between motor and compressor was checked and it was found well. Motor NDE and DE bearing clearances were found on higher side and minor scoring was observed on rollers (as shown in Figure 10.8).

Compressor pedestal bearing clearances were checked. No deviation was observed. Detailed inspection of motor stator was carried out. No abnormality was noticed (Figure 10.9).

10.4 MAINTENANCE WORK PERFORMED

Motor was de-coupled from the rest of the group drive. Before dismantling, initial alignment readings were taken for reference. Motor was taken out and the pulley was removed from shaft. Bearings of motor NDE and DE were removed (as shown in Figure 10.10). New bearing of motor NDE and DE were installed. Meanwhile, inspection and cleaning of rotor and stator was carried out. Finally, motor pulley was re-mounted on the shaft and assembled motor kept back in its original position. Motor trial was taken at full load and all vibration readings were found well within prescribed limit. Compressor boxed up (ready for operation) and made available for trial.

FIGURE 10.8 Damaged motor DE bearing.

FIGURE 10.9 Proactive checks for pedestal bearing, motor rotor and stator.

FIGURE 10.10 Motor DE bearing replacement work.

10.5 POST-MAINTENANCE OFFLINE VIBRATION READINGS

Table 10.1 represents the comparison of pre- and post-vibration data. With this data, it was noticed that, after replacing motor DE bearing, vibration readings were reduced to normal level which signified the successful execution of maintenance work.

10.6 POST-MAINTENANCE ONLINE VIBRATION READINGS

Online vibration reading of motor DE bearing was reduced from 5.8 to 0.8 mm/s (RMS) at full load condition (shown in Figure 10.11).

10.7 POST-MAINTENANCE VIBRATION SPECTRUM

From Figure 10.12, total magnitude was observed in the range of 0.7 (RMS) and 0.9 mm/s (RMS). The above vibration spectrum did not show any abnormality such as bearing damage frequency at 7740 cpm or looseness. Hence, it concluded the success of maintenance work execution.

TABLE 10.1
Vibration Report of SAC (Service Air Compressor)- 2B

Point		Position	Initial Readings 03/07/2018	After Replacing Motor Bearings 04/07/2018
			Velocity (mm/s) RMS	
Motor	(NDE)	H	**4.4**	**0.7**
		V	1.3	0.7
		A	**2.8**	**0.6**
	Acceleration – g	H	0.6	0.3
Motor	(DE)	H	**4.0**	**0.7**
		V	1.0	0.8
		A	**3.8**	**0.9**
	Acceleration – g	H	0.7	0.5

(Continued)

TABLE 10.1
Vibration Report of SAC (Service Air Compressor)- 2B

Point		Position	Initial Readings 03/07/2018	After Replacing Motor Bearings 04/07/2018
			Velocity (mm/s) RMS	
Compressor pedestal bearing	(DE)	H	0.8	0.9
		V	0.2	0.2
		A	1.2	1.5
	Acceleration – g	H	0.9	0.5
Compressor bearing	(NDE)	H	1.0	0.8
		V	1.0	0.7
		A	1.6	1.3
HP cylinder bearing	(DE)	H	2.7	2.5
		V	1.1	1.2
		A	3.6	3.6
HP cylinder bearing	(NDE)	H	2.7	2.8
		V	1.9	1.7
		A	5.6	3.0
LP cylinder bearing	(DE)	H	1.9	1.2
		V	1.2	1.5
		A	1.9	1.9
LP cylinder bearing	(NDE)	H	2.2	2.3
		V	1.5	1.8
		A	2.1	1.8

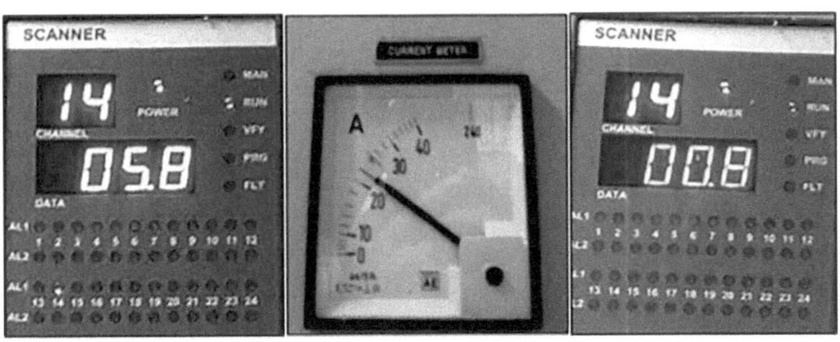

FIGURE 10.11 Online vibration reading of motor DE bearing for pre- and post-maintenance.

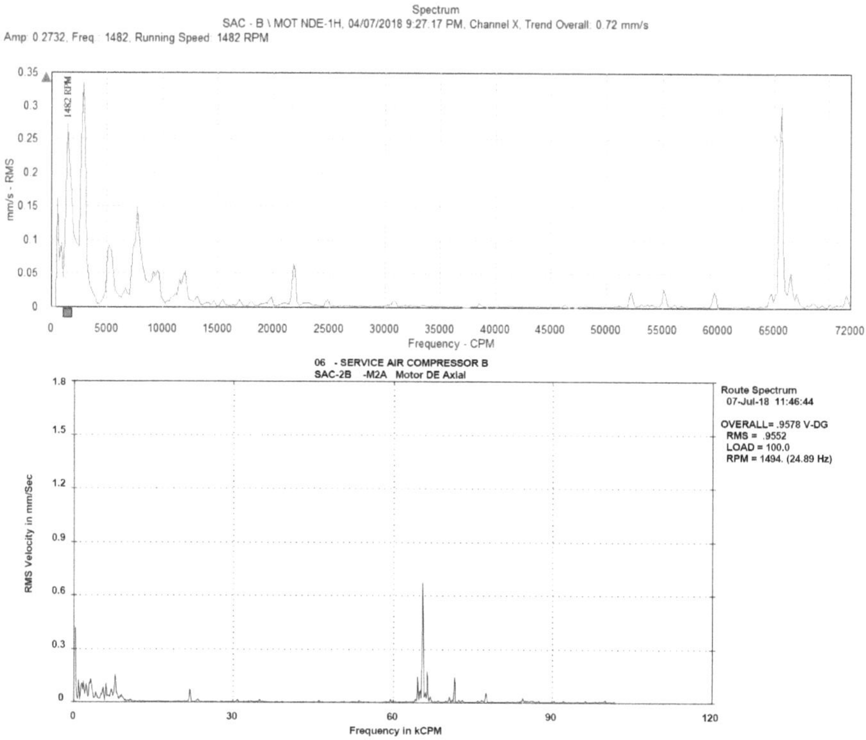

FIGURE 10.12 Spectrum of motor NDE bearing – (horizontal) and motor DE bearing (axial).

10.8 CONCLUSIONS

Compressed air is a secondary energy source which is required in many industries involving various processes. In thermal power stations, air compressors are provided for supplying compressed air for the operation of pneumatic drives, for atomizing of Light Diesel Oil (LDO) for proper combustion and for general cleaning purpose. Three Instrument Air Compressor (IAC) and two Service Air Compressor (SAC), two driers and four air receivers are provided for the above said purpose at ADTPS. Two IACs and one SAC are sufficient to meet instrument air and service air requirements. One IAC selected as 'Master' and another running compressor selected as 'Slave', third compressor is kept as standby. One SAC is running and other is kept as a standby. This air compressor system is common for both the units. Hence, any abnormality in this system affects the reliability of both the units. Therefore, looking at the criticality of air compressors the defect of this system needs to be identified at very early stage to avoid prolonged equipment downtime.

Presently, at ADTPS no additional motor is available for air compressor, and due to this, the importance or criticality of air compressors is enhanced. Due to any reason, if air pressure is reduced, standby air compressors pick up in auto and maintain air pressure as per the requirement for the functioning of pneumatic drives and for atomizing of LDO. Non-availability of stand-by air compressor during such situations may lead to reduction in overall instrument air pressure throughout the plant and will affect the reliability of both the units. This condition may lead to tripping of both the units and interruption in power generation, thereby affecting electric supply to the esteemed customers of Mumbai, India.

This case study has clearly demonstrated that condition-based vibration testing is a vital component in rotating machinery analysis and detection of abnormality at early stage. Vibration monitoring and analysis were helped in diagnosing the exact root cause of abnormity well in advance which can avoid the secondary damage to motor and compressor. Other major benefits include prevention of downtime, avoidance of major motor repair or replacement costs and reliability of unit.

11 Case Study 10
Coal Mill 2AB: Pinion Gear Reversal – February 2018

EXECUTIVE SUMMARY

The present case study is a success story of the implementation of Condition Monitoring techniques at ADTPS, wherein imminent Pinion Gear detoriation of Coal Mill – 2AB was diagnosed. Vibration monitoring and its analysis helped in diagnosing the exact root cause of abnormity of pinion gear problem at early stage and thus an initiation of secondary internal damages of girth gear (the gear with the pinion mesh) was prevented. This has saved plant from unplanned outage of critical auxiliary which are used for pulverizing of coal.

On 24 November 2015, the presentation was given by condition monitoring team in daily planning meeting to address fluctuating high vertical vibration problem of pinion NDE bearing due to deterioration of pinion gear. Moreover, it was decided to keep a close watch on vibration readings.

On 22 December 2015, 10 February 2016 and 7 December 2016, frequent misalignment was noticed due to backlash disturbance between pinion and girth gear. Also, a step noticed on each tooth of pinion gear during inspection. On 8 December 2016, pinion backlash adjusted, and mill was taken in service with stable high vibrations of 10.2–10.5 mm/s by deciding the limit value of 12.0 mm/s.

On 10 December 2016, high vibration issue of pinion NDE bearing was discussed with original equipment manufacturer. Pinion gear reversal was suggested and accordingly action was taken by the maintenance team of the mechanical department. From 11 December 2016 to 27 June 2017, pinion NDE bearing vertical vibration was found stable between 8.5 and 9.5 mm/s.

On 28 June 2017, once again alignment was carried out, and mill was taken in service with stable high pinion NDE bearing vertical vibration of 9.5 mm/s. On 14 August 2017, detailed vibration analysis along with OEM recommendations were discussed in daily planning meeting and decided to go for pinion reversal during the upcoming overhauling process.

From 6 January 2018 to 15 January 2018, coal mill pinion reversal and both pinion bearings replaced in the presence of OEM expert during annual overhaul conducted in January 2018. Girth gear and reducer gearbox inspection were carried out and no deviations were noticed.

11.1 PRE-MAINTENANCE HISTORY

On 24 November 2015, presentation was given by team Condition Monitoring Cell (CMC) in daily planning meeting to address fluctuating high vertical vibration problem of pinion of non-drive end (NDE) bearing due to deterioration of pinion

DOI: 10.1201/9781003432616-11

gear. It was decided to keep a close watch on online vibration readings. On 22 December 2015, 10 February 2016 and 07 December 2016, frequent misalignment was noticed due to backlash disturbance between pinion and girth gear (shown in Figure 11.1). Step was noticed on each tooth of pinion gear during inspection as shown in Figure 11.2.

On 8 February 2016, pinion backlash adjusted, and mill was taken in service with stable high vibrations of 10.2–10.5 mm/s by deciding a limit value of 12.0 mm/s for smooth changeover (Figure 11.3). On 10 February 2016, high vibration issue of pinion NDE bearing was discussed with OEM M/s BHEL Hyderabad. Pinion gear reversal was suggested and accordingly, action was taken by the team of mechanical maintenance department. Between 11 December 2016 to 27 June 2017, pinion NDE bearing vertical vibration was observed as stable in the range of 8.5–9.5 mm/s. On 28 June 2017, once again pinion NDE vertical vibration was increased as shown in Figure 11.4. Backlash was adjusted for proper meshing of pinion with girth gear and

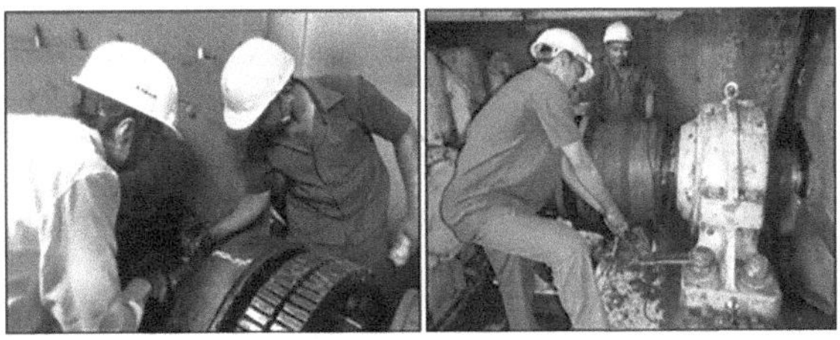

FIGURE 11.1 Alignment correction work.

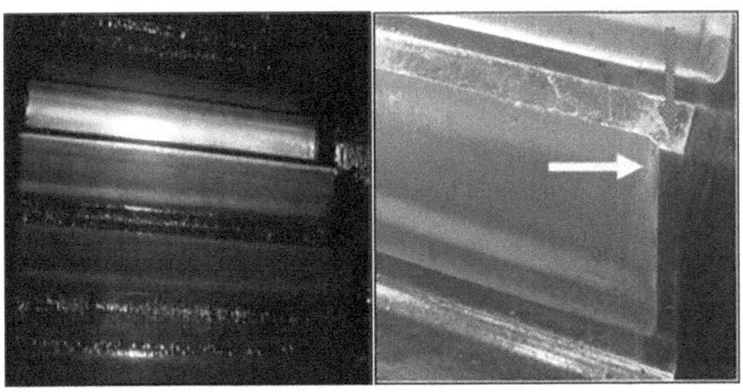

FIGURE 11.2 Step notice for pinion gear teeth.

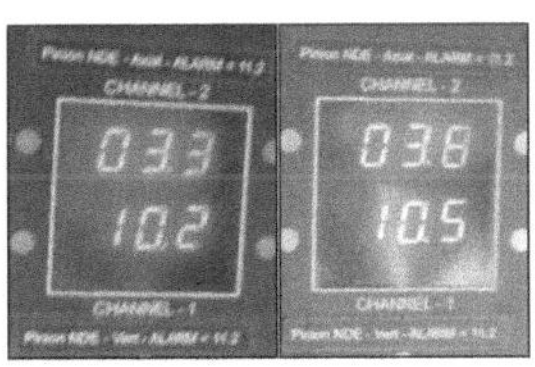

FIGURE 11.3 Online vibration reading of pinion bearing.

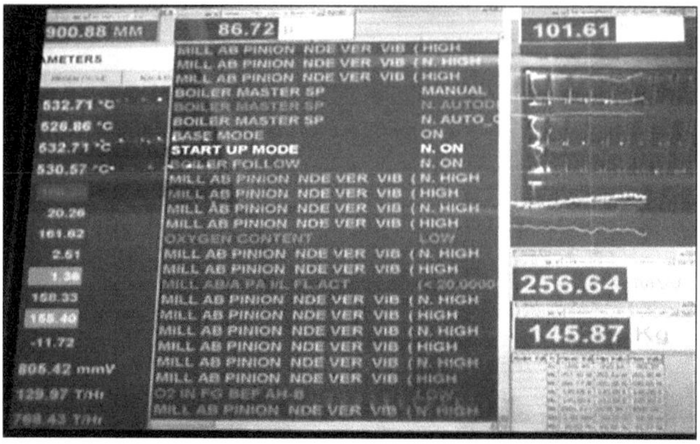

FIGURE 11.4 Alarm for high pinion bearing vertical vibration data (at control room).

FIGURE 11.5 Pinion backlash adjustment work.

with reference to that data alignment was carried out. Mill was taken in service with stable high vibration of 9.6 mm/s (Figure 11.5). On 14 August 2017, detailed vibration analysis along with OEM recommendations was discussed in daily planning meeting and decided to go for pinion reversal during upcoming overhaul. During annual overhaul period from 6 January 2018 to 15 January 2018, coal mill pinion reversal and both pinion bearings were replaced in presence of OEM expert during. Detailed inspection of girth gear and reducer gearbox was carried out. No deviations were noticed. On 5 February 2018, after annual overhaul completion, mill was taken in service. Vertical vibration of pinion NDE bearing was reduced from 10.5 to 2.4 mm/s with no fluctuations.

11.2 VIBRATION DIAGNOSIS

On 28 December 2017, vertical vibration of pinion NDE bearing was increased to 9.8 mm/s. Pinion gear mesh frequency was calculated and spectrum analysis for the root cause was depicted in Figure 11.6.

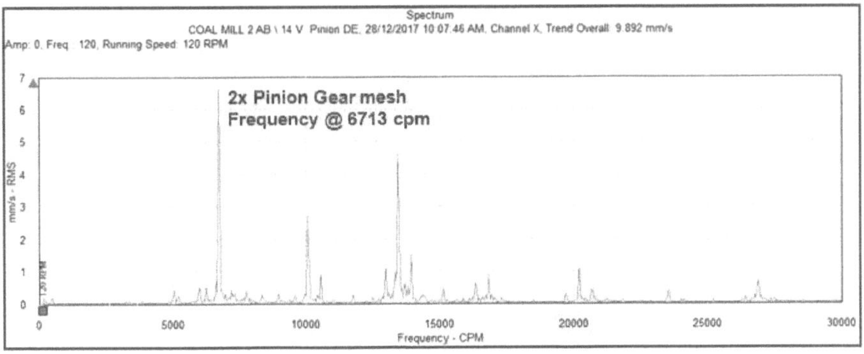

FIGURE 11.6 Vibration spectrum of pinion NDE bearing (vertical).

Calculation of gear mesh frequency:

- No. of gear teeth on pinion = 28
- Pinion RPM = 120
- Pinion gear mesh frequency = 28 × 120 = 3360
- 2x Pinion gear mesh frequency = 3360 × 2 = 6720 (Appox.)

11.3 FREQUENCY ANALYSIS

Pinion NDE bearing vertical vibration was maintained on higher side between 9.1 and 9.8 mm/s. Spectrum analysis revealed that there is increase in pinion gear mesh frequency of 6713 cpm which pinpoint to the existence of misalignment between pinion and girth gear. Normal bearing acceleration readings were eliminated existence of any bearing defect. Due to high pinion NDE bearing vibration fluctuations, it was decided to inspect pinion gear. Gear step was observed on each tooth of pinion gear. Thus, it was concluded that frequent misalignment between pinion and girth gear was found due to the deterioration of pinion gear teeth which results in high bearing vibration in vertical direction. Finally, the decision was taken for pinion gear reversal.

11.4 OBSERVATIONS DURING MAINTENANCE WORK

Alignment between pinion and girth gear was found to be disturbed as shown in Figure 11.7. Condition of both end coupling, pinion bearing housing and foundation bolts were found well. No scoring marks were observed on pinion bearings and shaft (Figure 11.8). Detailed inspection of girth gear was carried out in the presence of OEM. No scoring marks were observed (see Figure 11.9).

11.5 PERFORMED MAINTENANCE WORK

Pinion was de-coupled from the rest of the group drive (Figure 11.10). Panels of sound hood was required for removal of pinion. Girth gear greasing panel was removed. Arrangement was made for rotation of mill shell to remove the pinion. Complete

FIGURE 11.7 Initial alignment check.

FIGURE 11.8 Pinon gear and shaft inspection.

FIGURE 11.9 Girth gear inspection.

FIGURE 11.10 De-coupling of pinion gear.

FIGURE 11.11 Greasing panel (LHS) and pinion gear removal (RHS).

FIGURE 11.12 Pinion gear heating in progress.

pinion along with DE and NDE bearings and pinion coupling were removed as shown in Figure 11.11. Both pinion bearings and drive end (DE) side gear coupling were removed from shaft. Pinion gear was placed inside the fixture (Jugaad) made as per OEM drawing. Pinion gear was heated up to 150°C (as shown in Figure 11.12). With continuous heating, 200 tons pressure was applied with hydraulic jack and pinion gear was removed. Cleaning of pinion shaft and gear was carried out before the start of reversal work (as shown in Figure 11.13). Pinion gear was again heated in control way up to 160°C with the help of induction heating machine as shown in Figure 11.14. Shaft dimensions were recorded for reference and Molycoat (type of lubrication for smooth movement/travel of gear on shaft) was applied on pinion shaft just before installing the gear to minimize the friction as shown in Figure 11.15.

Pinion gear was installed on shaft in reverse direction, so that during operation unused face of pinion teethes would come in contact with girth gear as shown in Figure 11.16. Gear lock nut was tightened and kept pinion for natural cooling as shown in Figure 11.17. After cooling of pinion gear, both pinion new bearings (DE

FIGURE 11.13 Removal of pinion gear from shaft.

FIGURE 11.14 Heating of pinion gear to place on shaft in reverse direction.

FIGURE 11.15 Molycoat applied on pinion shaft.

FIGURE 11.16 Placement of pinion gear on shaft in reverse direction.

FIGURE 11.17 Lock nut tightening.

and NDE), adapter sleeve and gear coupling were mounted over the shaft. New pinion bearings clearances were checked and found within the prescribed limit. Assembled pinion is kept back in its original position as shown in Figure 11.18.

Alignment with respect to girth gear was carried out and backlash readings were considered as per the manual provided by OEM M/s BHEL Hyderabad (Figure 11.19). Keeping the pinion as a master, alignment of drive group was carried out (shown in Figure 11.20) and trial was taken at full load condition. Motor, reducer GB and pinion bearing vibrations were found within limit. Coal mill was boxed up (ready for operation) and made available to take into the service.

FIGURE 11.18 Installation of new pinion bearings.

FIGURE 11.19 Pinion gear backlash adjustment with girth gear.

FIGURE 11.20 Final alignment work.

11.6 POST-MAINTENANCE OFFLINE VIBRATION READINGS

On 5 February 2018, mill was taken in service after annual overhaul completion. Vertical vibration of pinion NDE bearing was finally reduced from 10.5 to 2.4 mm/s without any fluctuations.

Table 11.1 and Figures 11.21 and 11.22 represent the vibration database of pre- and post-maintenance. The overall reduction in the vibration level of pinion NDE bearing (vertical) confirmed the success of maintenance actions executed.

TABLE 11.1
Vibration Report of Coal Mill 2AB

			Initial Readings		After Pinion Gear Reversal	
			04/01/2018		05/02/2018	
Point		Position	Velocity (mm/s) RMS	Acceleration (g) Peak	Velocity (mm/s) RMS	Acceleration (g) Peak
Main motor	(NDE)	H	2.2	0.1	1.3	0.1
		V	2.1	-	1.1	-
		A	2.7	-	1.7	-
Main motor	(DE)	H	4.8	0.3	4.3	0.2
		V	4.6	-	3.8	-
		A	3.9	-	3.7	-
Reducer GB	(DE)	H	5.6	1.1	2.2	1.0
Input bearing		V	4.0	-	2.7	-
		A	4.7	-	3.7	-
Reducer GB	(NDE)	H	6.7	2.0	3.3	2.3
Input bearing		V	4.9	-	2.9	-
		A	5.4	-	3.9	-
Reducer GB	(NDE)	H	5.4	2.6	3.4	2.7
Intermediate		V	1.8	-	1.1	-
bearing		A	3.9	-	2.5	-
Reducer GB	(NDE)	H	7.1	0.8	3.7	0.6
Output		V	4.4	-	2.1	-
bearing		A	4.5	-	3.5	-
Pinion bearing	(DE)	H	5.1	1.2	2.4	0.5
		V	**6.5**	-	**2.3**	-
		A	4.2	-	1.6	-
Pinion bearing	(NDE)	H	3.4	1.0	2.1	0.4
		V	**9.8**	-	**2.1**	-
		A	5.3	-	1.6	-

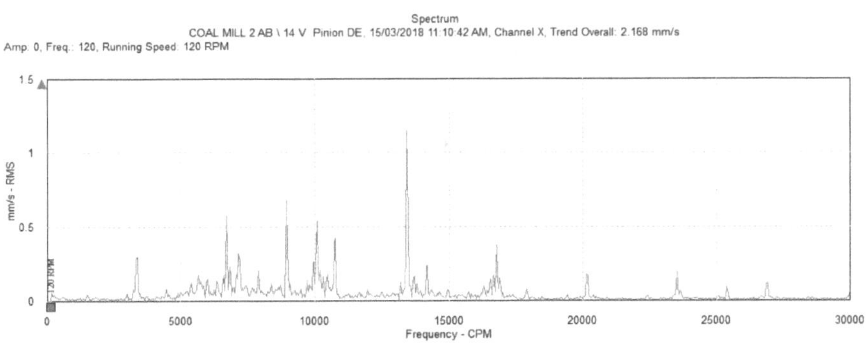

FIGURE 11.21 Post-maintenance vibration spectrum.

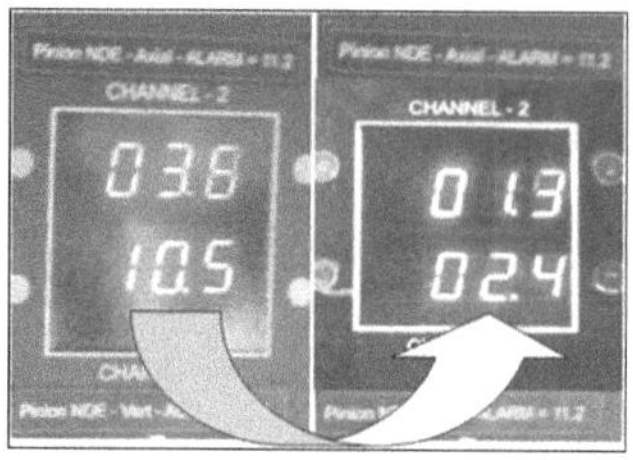

FIGURE 11.22 Pre- and post-maintenance online vibration readings.

11.7 POST-MAINTENANCE VIBRATION SPECTRUMS

Total Magnitude: 2.1 mm/s (RMS)

The above vibration spectrum did not show any abnormality such as misalignment or looseness.

11.8 POST-MAINTENANCE ONLINE VIBRATION READINGS

Figure 11.22 summarises the vibration reading of pre- and post-maintenance.

11.9 CONCLUSIONS

Coal mill is an important auxiliary used to pulverise coal in required size for efficient combustion in boiler. Coal mill supplies coal as fuel to boiler, the reliability of coal mill needs to be very high. It is a critical auxiliary in the process because its failure directly affects the generation loss of 0.125 MUs/h which costs revenue of around Rs. 437,500 per hour. There is total three nos. of coal mills (i.e., two are running at full load and other is in stand-by condition). This provision is made because coal mill is critical equipment in the system and if it fails during the non-availability of stand-by mill, it may lead to tripping of plant and interruption in power supply to our valued customers of Mumbai, India.

The present case study has clearly demonstrated that condition-based vibration testing is a vital component in rotating machinery analysis and detection of abnormality in advance. Vibration monitoring and analysis helped in diagnosing the exact root cause of abnormity of pinion gear problem at early stage and thus avoided the initiation of secondary internal damages of girth gear and bearings. This pinion reversal work execution has saved the plant from an unplanned outage of critical auxiliary used for pulverizing of coal and spares associated maintenance cost. Also, other major benefit includes prevention of downtime, avoidance of major gearbox repair or replacement costs.

12 Case Study 11 *ID Fan 2A: Hydraulic Coupling Bearing Deterioration – October 2017*

EXECUTIVE SUMMARY

The present case study is a success story of the implementation of condition monitoring techniques at ADTPS, wherein imminent hydraulic coupling (HC) bearing deterioration of HT auxiliary (ID Fan – 2A) was diagnosed. This case study has clearly demonstrated that acceleration is a valuable parameter in rotating machinery analysis and is especially useful in detecting early and advanced bearing degradation even vibration levels are on the lower side at initial deterioration period of bearing. Oil wear debris analysis (WDA) and ultrasonic measurement played a major supportive role in confirming the damage of HC bearing. It is a highly powerful diagnostic tool and helps in identifying the root cause leading to wear out of components and guided plant engineer to plan maintenance work in case of any emergency.

On 28 January 2017, the acceleration value of HC bearing deviated from 2.4 to 3.1 g without deviation in vibration level. It was decided to send an oil sample for WDA.

On 14 February 2017, WDA oil report confirmed the presence of medium alloy bearing wear particles of size ranging up to 68 μm. It was decided to keep a close watch on all condition monitoring parameters.

On 30 May 2017, the acceleration value of HC input DE bearing was increased from 3.1 to 3.8 g. Also, the ultrasonic value was deviated from 3 to 6 LED lights. This shows severity indicator screen on the ultrasonic data collector instrument.

On 14 July 2017, vibration and acceleration monitoring system was installed for HC input DE bearing and the local operator was educated regarding all limit values. Also, HC oil sample was sent for WDA On 9 August 2017, the acceleration value of HC input DE bearing was increased from 3.8 to 4.2 g. Detailed vibration analysis, WDA and ultrasonic report along with various probable causes discussed in daily HOD meeting and decided to replace HC during unit shutdown.

On 20 September 2017, at 190 MW plant load condition, 0.7–1.0 mm/s deviation in horizontal vibration was observed in motor, HC and fan. On 21 September 2017, at 250 MW plant load condition, horizontal vibration for HC and Fan was observed on higher side i.e., 2.9 and 7.7 mm/s.

12.1 PRE-MAINTENANCE HISTORY

On 28 January 2017, the acceleration value of HC input bearing was deviated from 2.4 to 3.1 g without any deviation observed in vibration level (Figure 12.1).

DOI: 10.1201/9781003432616-12

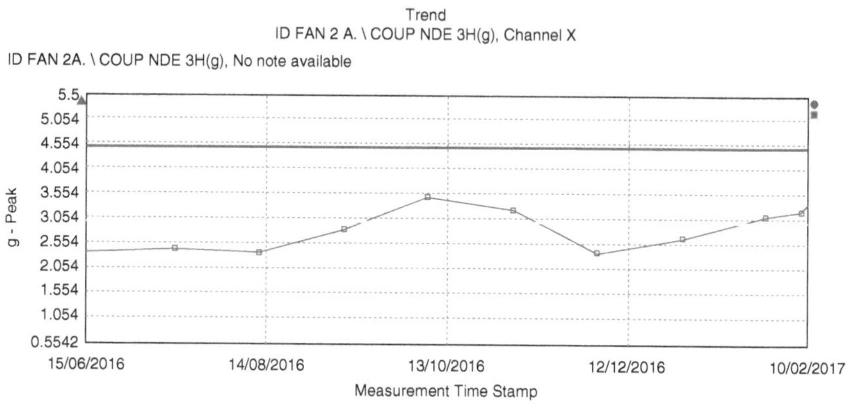

FIGURE 12.1 Acceleration trend for hydraulic coupling (HC) input bearing.

Other ID Fans	Acceleration HC DE (g)	Acceleration HC NDE (g)
1A	0.9	0.7
1B	1.9	1.1
2B	1.4	1.0

FIGURE 12.2 Acceleration value comparison for hydraulic coupling (HC) input bearing.

In addition, no deviation was observed in any process parameters of HC. Acceleration value was compared with other ID fans and deviations were observed in ID fan 2A only (Figure 12.2).

It was decided to send an oil sample for wear debris analysis (WDA). On 14 February 2017, WDA oil report confirmed the presence of medium alloy bearing wear particles of size ranging up to 68 μm (Figure 12.3). WDA oil report was compared with other ID fans and deviations were observed in ID fan 2A only (Figure 12.4). It was decided to keep a close watch on all condition monitoring parameters. On 30 May 2017, the acceleration value of HC input bearing increased from 3.1 to 3.8 g. 9. Also, the ultrasonic value was deviated from 3 to 6 LEDs (as shown in Figure 12.5). On 14 July 2017, online vibration and acceleration monitoring system were installed for HC input DE bearing and the local operator was educated regarding all limit values (Figure 12.6). Also, HC oil sample was sent for WDA. On 9 August 2017, the acceleration value of HC input bearing was increased from 3.8 to 4.2 g. Detailed vibration analysis, WDA and ultrasonic report along with various probable causes were discussed in HoD meeting and decided to replace HC during opportunity. On 20 September 2017, motor, HC and fan horizontal vibration was found to be deviated

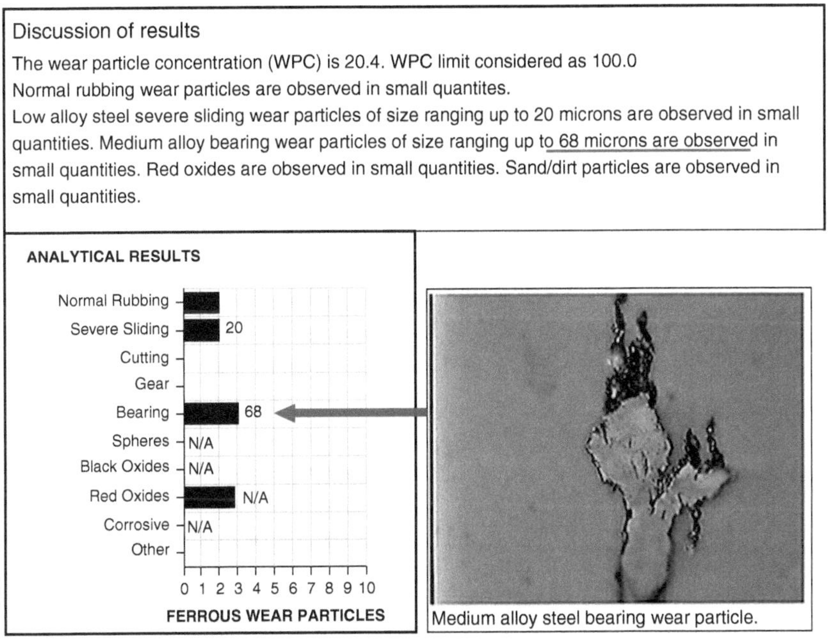

FIGURE 12.3 Wear debris oil analysis report interpretation.

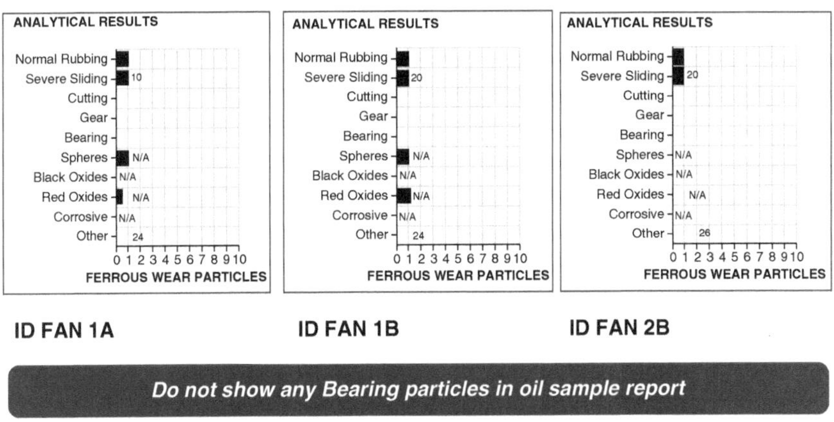

FIGURE 12.4 Wear debris oil analysis report comparison with other ID fans.

by 0.7–1.0 mm/s at 190 MW. On 21 September 2017, HC and fan online horizontal vibration reading was recorded on higher side (i.e., 2.9 and 7.7 mm/s) at 250 MW load (see Figure 12.7). On 22 September 2017, in view of increasing trend of vibration, acceleration, ultrasonic and WDA value, it was decided to replace HC during unit shutdown.

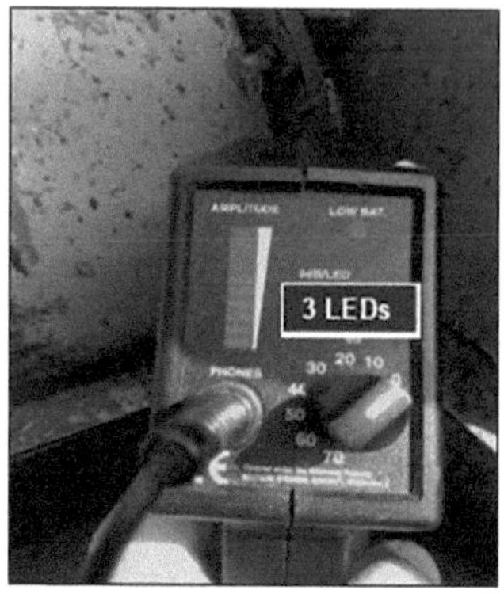

FIGURE 12.5 Ultrasound data collection at hydraulic coupling (HC) input bearing.

FIGURE 12.6 Online vibration and acceleration value of hydraulic coupling (HC) input bearing.

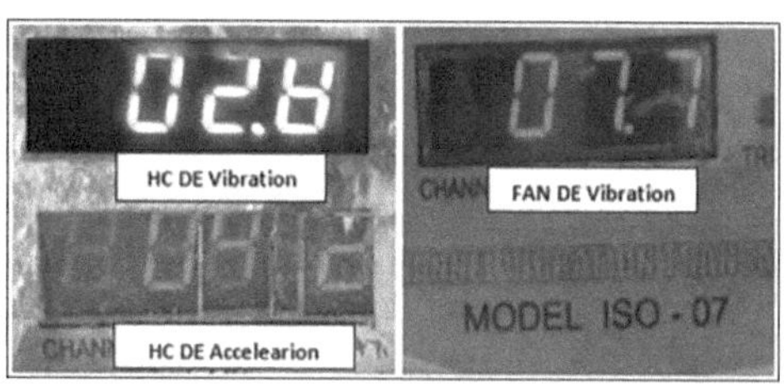

FIGURE 12.7 Online hydraulic coupling (HC) input bearing and fan vibration.

12.2 VIBRATION DIAGNOSIS

On 21 September 2017, acceleration value of HC input DE bearing was reached to 4.2 g. The spectrum analysis for root cause is investigated as follows (Figure 12.8):

Non-synchronous value: 2.3814; Synchronous value: 1.5431; Sub-synchronous value: 0.0016.

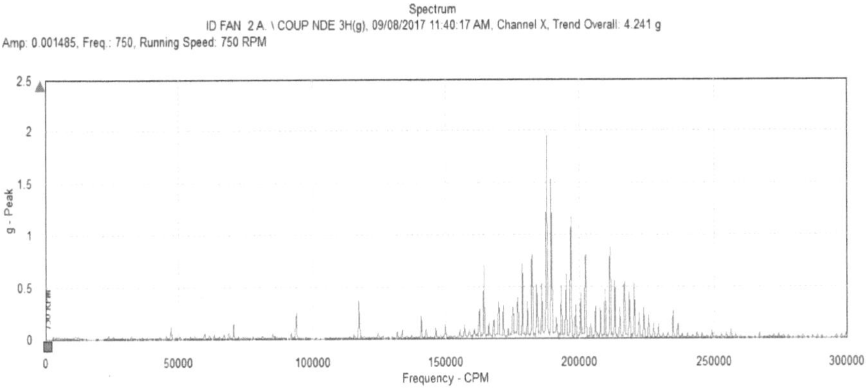

FIGURE 12.8 Acceleration spectrum of hydraulic coupling input bearing.

It was found that there was an increase in acceleration value of HC input DE bearing with high horizontal vibrations. Spectrum analysis revealed that it was observed that "g" value was high, and these high frequencies got excited. These high frequencies were observed due to bearing defect. To confirm the damage of bearing, WDA of oil and ultrasonic analysis were carried out.

12.3 WDA OF OIL

Routine oil condition monitoring (OCM) identifies small problems within machinery components before they become catastrophic failures which can incur costly repairs and bring production to a halt. WDA complements vibration analysis by providing independent and advanced indications of mechanical wear. It is a technique that is based on the systematic collection of oil samples from oil-lubricated machines. This method identifies, isolates and classifies wear particles from machine parts. A magnetic field is used to sort the wear particles in the flowing oil. This analysis is microscopic evaluation of wear particles which identifies the type and severity of wear and generation mode that helps in understanding the source of wear component. It also helps in identifying the root cause leading to the wear out of components.

WDA oil report of HC coupling confirmed the presence of medium alloy bearing wear particles of size up to 68 µm as shown in Figure 12.9.

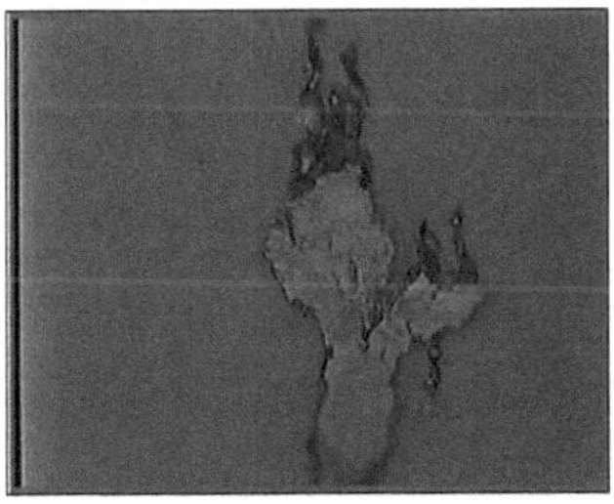

FIGURE 12.9 Bearing wear particle found in oil analysis.

12.4 ULTRASONIC MEASUREMENT AND ANALYSIS

Ultrasonic is the application of ultrasound. Ultrasound is used for the detection and measurement of high frequencies. Ultrasound is an oscillating sound pressure wave with a high frequency. These high frequencies typically range from 20 to 100 KHz and beyond the range of human hearing. Ultrasound instruments receive high-frequency sounds produced by operating equipment, electrical emissions and leaks. The instruments electronically translate ultrasound frequencies down into the audible range. These frequencies are heard through headphones and observed as intensity on a display panel.

With reference to Figures 12.10 and 12.11, deviation in oil parameter and high ultrasonic reading) confirmed that HC bearing deterioration was in progress.

12.5 FREQUENCY ANALYSIS

Acceleration value of HC input bearing was on higher side (4.2 g) with high vibration. Normally, it ranges between 1.5 and 2.0 g at full load condition. It was suspected that HC input deterioration was in progress. Acceleration spectrum analysis (Figure 12.8) revealed that there was an increase in higher frequencies and the contribution of non-synchronous energy (any vibration that is greater than the 1× frequency but is not a whole number multiple of it) was found at higher side which was mainly due to bearing defects. These high-frequency stress waves' events occur at repeatable time intervals corresponding to a bearing defect frequency, then it is likely that a

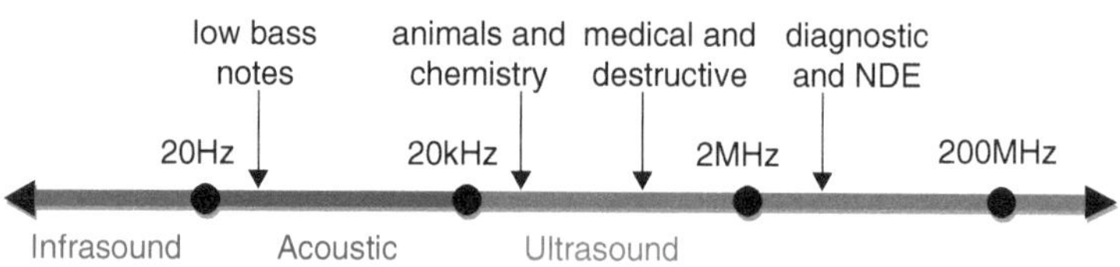

FIGURE 12.10 Fundamentals of ultrasound.

FIGURE 12.11 Deviation noticed in ultrasound data at hydraulic coupling (HC) input bearing.

bearing defect was present. These frequency stress waves' event occur at repeatable time intervals corresponding to a bearing defect frequency, then it proved that bearing defect was present. Stress waves are high frequency vibrations, caused due to: Impact of two metal surfaces with each other; Passing of bearing rolling element over any defect; due to fatigue cracking, excessive friction, and abrasive wear. Stress waves are short in duration, microseconds for fatigue cracking, milliseconds for metal-to-metal impact.

It was concluded that the acceleration value of HC input bearing increased due to early-stage deterioration of bearings which in turn increased the ultrasonic level and deviation in oil parameters. It was decided to replace HC during unit shutdown opportunity.

12.6 OBSERVATIONS DURING MAINTENANCE WORK

On 22 September 2017, in view of increasing the trend of vibration, acceleration, ultrasonic and WDA values, it was decided to replace the hydraulic coupling (HC) during unit shutdown. During maintenance, it was observed that clearances of HC input and output bearing are on higher side. Inspection of both fan bearings along with fan internal components was carried out. No other abnormality or scoring marks were observed during this inspection.

Alignment between HC and fan was found to be disturbed (see Figure 12.12). HC bearing clearances was observed on higher side. No scoring marks were observed on motor and fan shaft as shown in Figure 12.13. Detailed inspection of both fan DE and NDE bearing were carried out and no further deviation was observed in bearing clearances (Figure 12.14). Fan internal inspection was done and no abnormality was noticed as shown in Figure 12.15.

12.7 PERFORMED MAINTENANCE WORK

HC was de-coupled from the rest of the group drive (see Figure 12.16). Before dismantling, initial alignment readings were taken for reference. HC oil draining was carried out as shown in Figure 12.17. HC was removed and disc pack coupling set was installed to new HC, motor and fan shafts. New HC was placed as shown in Figure 12.18. HC was aligned with motor and fan shown in Figure 12.19.

FIGURE 12.12 Initial alignment checkby maintenance team.

FIGURE 12.13 Observations for motor and fan shaft during maintenance work.

FIGURE 12.14 Fan NDE and DE bearing inspection.

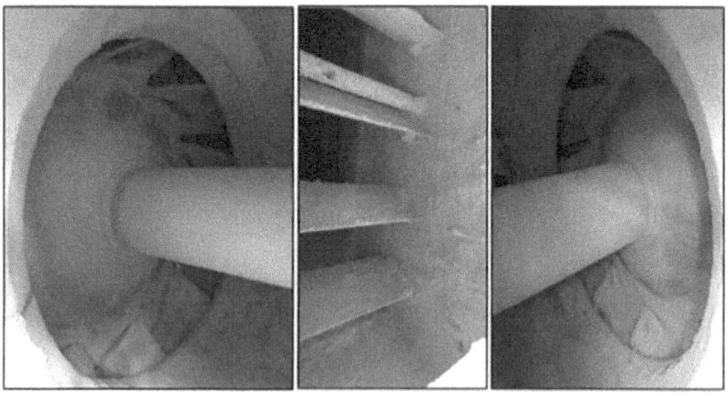

FIGURE 12.15 Fan impeller and shaft inspection.

FIGURE 12.16 De-coupling of hydraulic coupling.

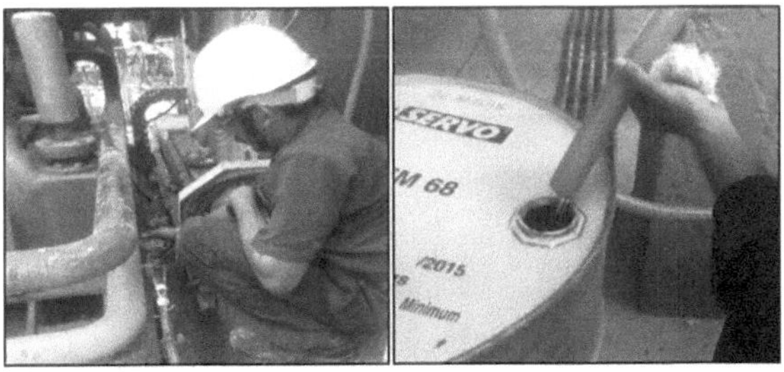

FIGURE 12.17 Hydraulic coupling (HC) oil draining process.

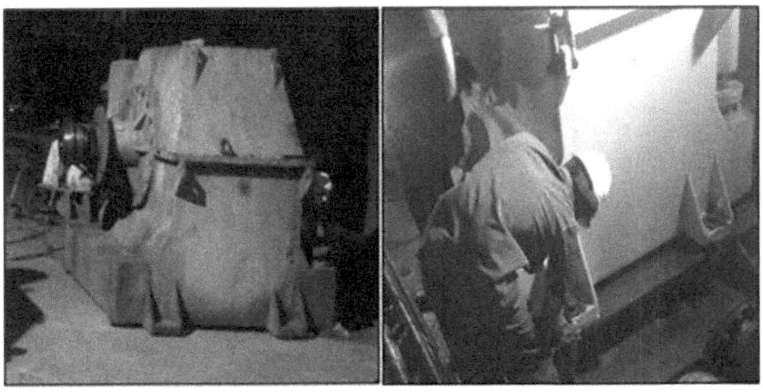

FIGURE 12.18 New hydraulic coupling (RHS) replaced against older one (LHS).

Fresh oil filling was done to new HC as shown in Figure 12.20. Final couplings between motor and HC, and HC and fan were carried out as shown in Figure 12.21. Trial was taken at full load and all vibration, acceleration and ultrasonic readings were found within limit. ID fan boxed up (ready for operation) and made available for taking in service.

FIGURE 12.19 Alignment of new hydraulic coupling with motor and fan.

FIGURE 12.20 Fresh oil filling to new hydraulic coupling.

FIGURE 12.21 Final coupling done after alignment work.

12.8 POST-MAINTENANCE OFFLINE VIBRATION READINGS AND ULTRASONIC LEVEL

Table 12.1 and Figures 12.22 and 12.23 represent the comparison of pre and post online and offline vibration, acceleration data with spectrum and ultrasonic data. All readings reduced to normal values after the replacement of HC. It was concluded that the successful execution of maintenance activities was performed by mechanical team.

TABLE 12.1
Vibration Report of ID Fan 2A

			Initial Readings		After Replacing Hydraulic Coupling	
			21/09/2017		10/10/2017	
		Position	Velocity (mm/s) RMS	Acceleration (g) Peak	Velocity (mm/s) RMS	Acceleration (g) Peak
Motor	(NDE)	H	1.2	1.4	0.8	1.2
		V	1.6	-	1.4	-
		A	0.6	-	1.0	-
Motor	(DE)	H	1.2	2.9	0.7	2.5
		V	0.6	-	0.6	-
		A	0.8	-	0.9	-
Hydraulic coupling	(DE)	H	**2.7**	**4.2**	**1.6**	**1.2**
		V	1.2	-	1.1	-
Input bearing		A	2.5	-	2.6	-
Hydraulic coupling	(NDE)	H	1.7	**1.9**	1.5	**0.5**
		V	1.6	-	1.4	-
Output bearing		A	2.7	-	2.5	-
Fan bearing	(DE)	H	0.9	-	0.4	-
		V	0.6	-	0.9	-
		A	0.4	-	0.3	-
Fan bearing	(NDE)	H	1.4	-	0.7	-
		V	1.6	-	0.8	-
		A	0.9	-	0.7	-
Ultrasonic level			**6 LEDs**		**3 LEDs**	
Online vibration and acceleration						

12.9 POST-MAINTENANCE ACCELERATION SPECTRUMS

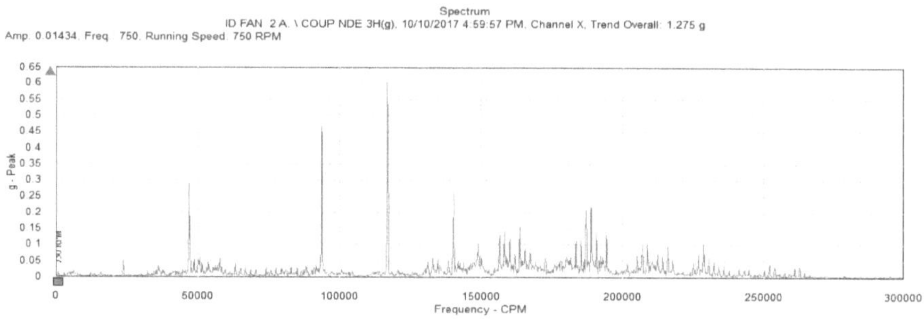

FIGURE 12.22 Reduction in hydraulic coupling input bearing acceleration value to 1.2 g.

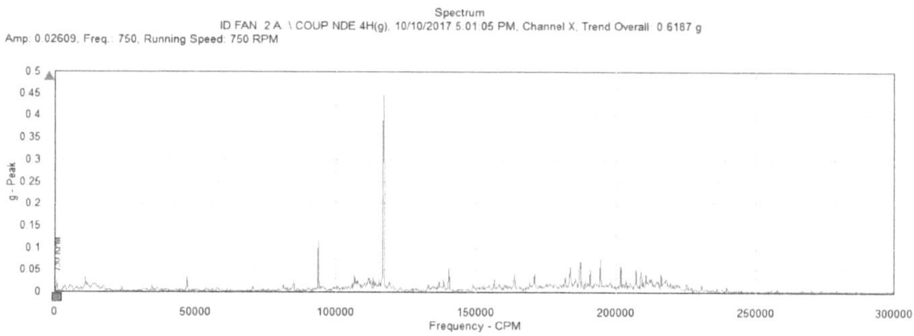

FIGURE 12.23 Reduction in hydraulic coupling output bearing acceleration value to 0.6 g.

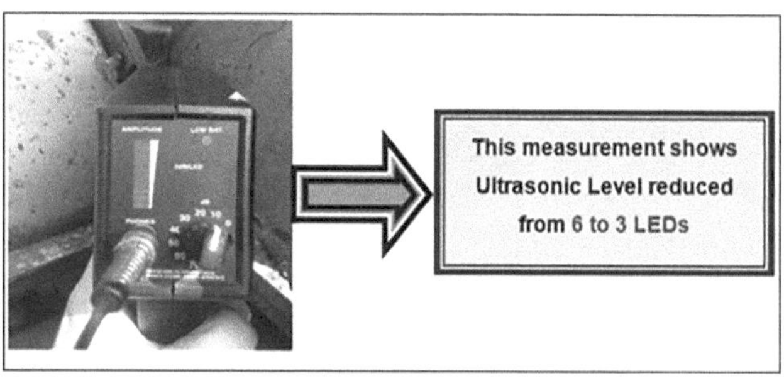

FIGURE 12.24 Post-maintenance ultrasound data.

12.10 POST-MAINTENANCE ULTRASONIC READINGS

Post-maintenance acceleration spectrum and ultrasonic level (depicted in Figures 12.22–12.24) do not show any abnormality such as bearing damage or looseness.

12.11 CONCLUSIONS

Induced draft fan is a critical equipment which is used to extract the flue gases out of furnaces via economiser, Air Pre-Heater (APH) through stack to atmosphere. It assists the FD fan by drawing out combustible gases from the furnace and maintains the partial negative pressure inside the Boiler.

Since ID fan has no standby, the reliability of ID fan needs to be very high. Boiler is provided with two nos. of ID fans. Each ID fan is provided with regulating damper control and scoop control for controlling the loading on fans. This made ID fan a critical equipment in the system. If any fan motor fails during operation, it will lead to generation loss and in extreme case tripping of plant and interruption in power supply occurs to our valued customers in Mumbai, India.

The present case study has clearly demonstrated that acceleration is a valuable tool in rotating machinery analysis and is especially useful in detecting early and advanced bearing degradation even when the vibration levels are on the lower side. Wear debris oil analysis and ultrasonic measurement played a major supportive role in confirming the damage of HC bearing. It is a very powerful diagnostic tool. Each vibration analyst should use to improve vibration analysis accuracy as this can help the mechanical maintenance team to become more proactive and accurate.

The secondary damage to motor and fan bearing is avoided using these techniques and also other major benefit includes prevention of downtime, avoidance of major motor repair or replacement costs. The application of co-ordinated condition monitoring helped in diagnosing the exact root cause of abnormity well in advance and guided plant engineer to plan maintenance work during the nearest available opportunity with no generation loss.

13 Case Study 12
Coal Conveyor 4A (CHP): Gearbox Input Bearing Damage – August 2017

EXECUTIVE SUMMARY

Following is a success story of the implementation of condition monitoring techniques at ADTPS, wherein imminent Reducer Gearbox Input bearing damage of HT auxiliary Coal Conveyor – 4A was diagnosed. This case study has clearly demonstrated that condition-based vibration testing is a vital tool in rotating machinery analysis and detection of abnormality at early stage. Chemical analysis and ultrasonic measurement acted as a major supportive role to confirm the damage of gearbox input bearing. It is a very powerful diagnostic tool. Every vibration analyst should use these supportive techniques to improve vibration analysis accuracy as this can help the maintenance department to become more proactive and accurate.

On 29 July 2017, local operator noticed fluctuations in vibrations (online data) of gearbox input bearing. After collecting the offline data, fluctuations were observed in horizontal vibration of reducer input DE bearing between 1.6 and 4.3 mm/s (RMS) and the acceleration value was deviated from 3.0 to 3.5 g. Moreover, ultrasonic reading was deviated from 4 to 8 LEDs. There was no other deviation observed in reducer gearbox process parameters. From vibration spectrum analysis and high ultrasonic reading, it was suspected that main motor DE bearing deterioration was in progress. Meanwhile, oil sample was sent for chemical analysis.

On 31 July 2017, one more set of vibration reading was taken and observed that fluctuations increased from 4.3 to 6.8 mm/s (RMS) and the acceleration value also increased from 3.5 to 4.4 g. Chemical analysis report confirmed the deviation in oil parameters, too. In view of increasing trend of all condition monitoring parameters, it was decided to stop coal conveyor for replacement of reducer input DE bearing.

During maintenance, it was observed that reducer input DE bearing clearances were on higher side and bearing found damaged. Inspection of intermediate and output bearing along with gear teethes were carried out. No scoring marks were observed. This practice has saved the plant from huge losses in terms of unplanned breakdown time and associated maintenance costs.

13.1 PRE-MAINTENANCE HISTORY

On 29 July 2017, local operator noticed fluctuations in online vibration readings of gearbox input bearing. After collecting the offline data, fluctuations in horizontal vibration (between 1.6 and 4.3 mm/s) were observed in reducer input DE bearing (as shown in Figure 13.1). In addition, acceleration value was deviated from 3.0 to 3.5 g (see Figure 13.2). Ultrasonic reading was deviated from 4 to 8 LEDs. However, no

DOI: 10.1201/9781003432616-13

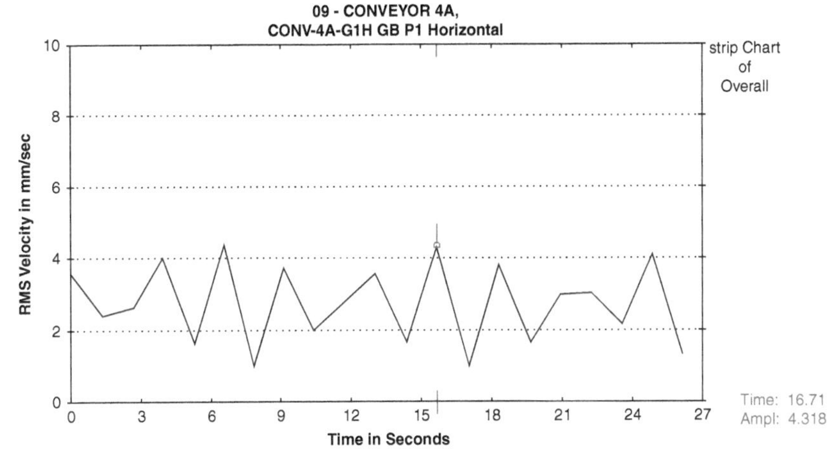

FIGURE 13.1 Offline vibration data trend for gearbox input bearing.

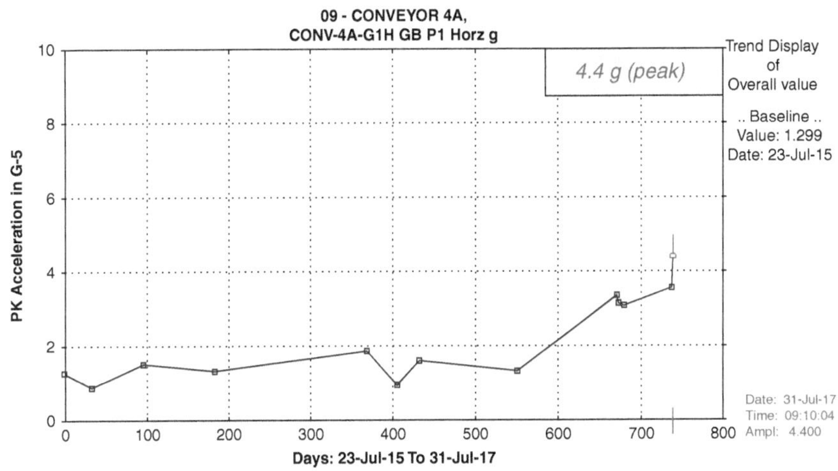

FIGURE 13.2 Acceleration value trend for gearbox input bearing.

deviation was observed in other process parameters of reducer gearbox. Meanwhile, oil sample was sent for chemical analysis. High vibration issue along with probable causes were discussed in daily planning meeting. It was decided to keep a close watch on online vibration readings reducer input DE bearing (Figure 13.3). On 31 July 2017, one more set of vibration reading was taken and observed that fluctuations were increased from 4.3 to 6.8 mm/s (RMS) (Figure 13.4). Acceleration value was also increased from 3.5 to 4.4 g. Chemical analysis report confirmed the deviation in oil parameters (Figure 13.5). In view of the increasing trend of all condition monitoring parameters, it was decided to stop coal conveyor for replacement of reducer input DE bearing.

FIGURE 13.3 Ultrasound data collection at gearbox input bearing.

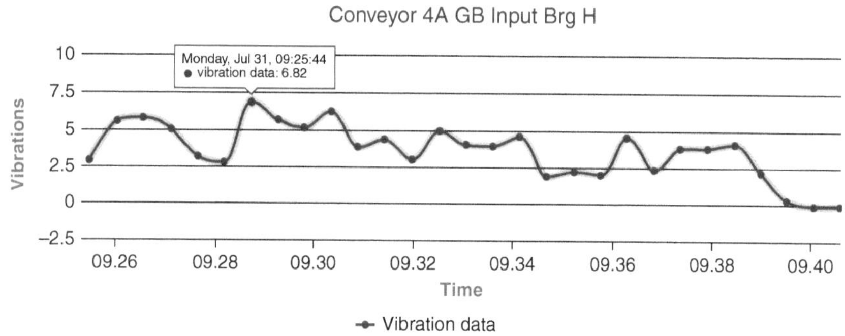

FIGURE 13.4 Online vibration data trend for gearbox input bearing.

13.2 VIBRATION DIAGNOSIS

On 31 July 2017, the acceleration value of reducer input DE bearing reached to 4.4 g. The spectrum analysis for root cause is investigated as follows: Observed the following readings with reference to Figure 13.6: Non-synchronous value: 3.674; Synchronous value: 2.543; Sub-synchronous value: 0.001. It was found that there was an increase in acceleration value of reducer input DE bearing with fluctuations in horizontal vibrations. Spectrum analysis revealed that "g" value was high, and these high frequencies got excited. These high frequencies were observed due to bearing defect. To confirm the damage of bearing, ultrasonic analysis was carried out.

13.3 ULTRASONIC MEASUREMENT AND ANALYSIS

Ultrasonic is the application of ultrasound. Ultrasound is used for detection and measurement of high frequencies. Ultrasound is an oscillating sound pressure wave with a high frequency. These high frequencies typically range from 20 to 100 kHz and beyond the range of human hearing. Ultrasound instruments receive high-frequency sounds produced by operating equipment, electrical emissions, and leaks. The instruments electronically translate ultrasound frequencies down into the audible range. These frequencies are heard through headphones and observed as intensity on a display panel shown in Figure 13.7.

The above analysis shows ultrasonic reading was on higher side (eight LEDs), which confirmed that deterioration of reducer input DE bearing was in progress.

DAHANU THERMAL POWER STATION

CHEMICAL DEPARTMENT

Servo mesh SP 320

Sampling Date: 31/07/2017 Reporting Date: 31/07/2017

Sr. No.	PARAMETERS	Conveyor # 4A
1.	Appearance	Thick, Dark Black
2.	Density gm/cc at 15°C	0.9237
3.	Moisture in ppm	41.95
4.	Viscosity in c St at 40°C	418
5.	Impurities %	0.189

Note: **1. Viscosity is on higher side**
2. Coal Dust observed

LIMITS:

Parameters	Servo mesh SP 320
Moisture in ppm	1000
Viscosity in c St at 40°C	288 to 352
Impurities (Max.) %	0.50

FIGURE 13.5 Chemical analysis report of gearbox oil.

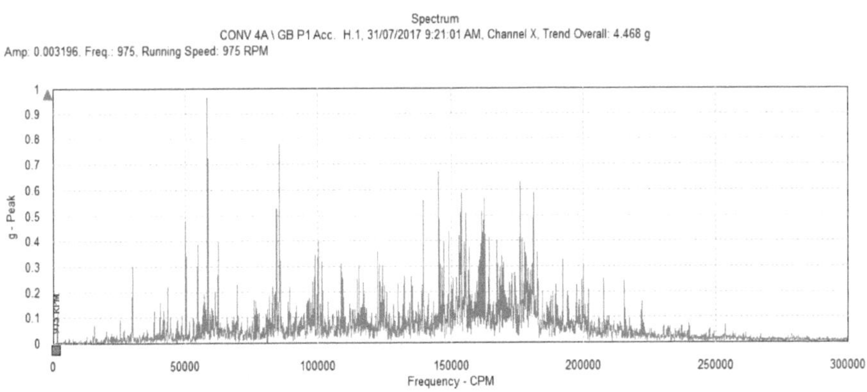

FIGURE 13.6 Acceleration spectrum of reducer input DE bearing.

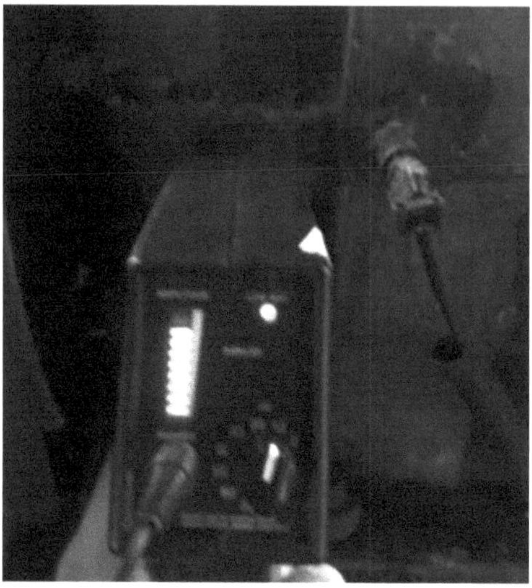

FIGURE 13.7 Fundamentals of Ultrasound.

13.4 FREQUENCY ANALYSIS

Acceleration value of reducer input DE bearing was on higher side (4.4 g) with fluctuation in vibration. Normally, it ranges between 1.5 and 2.0 g at load condition. It was suspected that reducer input DE bearing deterioration was in progress. Acceleration spectrum analysis (represented in Figure 13.6) revealed that there was an increase in higher frequencies and the contribution of non-synchronous energy (Any vibration that is greater than the 1× frequency but is not a whole number multiple of it) was at higher side which was mainly due to bearing defects. These high frequency stress waves' events occurred at repeatable time intervals corresponding to a bearing defect frequency, then it proved that a bearing defect was present. Stress waves are high-frequency vibrations, caused due to: Impact of two metal surfaces with each other; passing of bearing rolling element over any defect; fatigue cracking, excessive friction and abrasive wear. Stress waves are short in duration, microseconds for fatigue cracking, milliseconds for metal-to-metal impact. It was concluded that the acceleration value of reducer input DE bearing increased due to early-stage deterioration of bearings which in turn increased the ultrasonic level and deviation in oil parameters (refer to Figures 13.5 and 13.7). It was decided to stop coal conveyor for replacement of reducer input DE bearing.

13.5 OBSERVATIONS RECORDED DURING MAINTENANCE WORK

Condition of fluid coupling and reducer GB cooling fan was found well and ok. Reducer GB input DE bearing sitting area (bearing rest area) was found good in condition (Figure 13.8). Reducer GB input DE bearing was found loose as shown in Figure 13.9. Reducer GB input DE bearing clearances were found on higher side, and minor scoring marks were observed on rollers as shown in Figure 13.10.

FIGURE 13.8 Gearbox input DE bearing sitting area (bearing rest area).

FIGURE 13.9 Loose input bearing on shaft.

FIGURE 13.10 Scoring marks on input bearing rollers.

Detailed inspection of all bearings of intermediate and output shafts was carried out. No deviations were observed. In addition, detailed inspection of input, intermediate and output gears was carried out. No scoring marks were observed (see Figure 13.11).

13.6 PERFORMED MAINTENANCE WORK

Coal conveyor motor was de-coupled from rest of the group drives. Before dismantling, initial alignment readings were taken for reference. Reducer gearbox input DE bearing was removed and replaced with a new one as shown in Figure 13.12. After internal cleaning of reducer GB, oil flushing was done, and old oil was replaced with fresh oil (100L). Reducer GB top cover was placed, and motor was re-aligned with rest of drives as shown in Figure 13.13. Trial of coal conveyor was taken at full load and all vibrations and ultrasonic readings were found within the prescribed limit. Coal conveyor boxed up (ready for operation) and made it available for taking in service.

FIGURE 13.11 Inspection of gears by mechanical maintenance team.

FIGURE 13.12 Reducer gearbox input bearing replacement.

FIGURE 13.13 Gearbox boxed up along with its final alignment.

TABLE 13.1
Vibration and Ultrasonic Report of Coal Conveyor 4A

			Initial Readings 31/07/2017		After Replacing Reducer Input DE Bearing 02/08/2017	
		Position	Velocity (mm/s) RMS	Acceleration (g) Peak	Velocity (mm/s) RMS	Acceleration (g) Peak
Motor	(NDE)	H	3.5	1.4	2.2	1.2
		V	2.5	-	1.1	-
		A	1.0	-	0.5	-
Motor	(DE)	H	3.1	2.9	1.5	2.5
		V	1.4	-	0.9	-
		A	1.1	-	0.4	-
Reducer GB Input bearing	(DE)	H	**4.6**	**4.4**	**3.1**	**1.9**
		V	1.9	-	1.2	-
		A	1.5	-	1.2	-
Reducer GB Intr. bearing	(NDE)	H	1.3	**5.4**	0.7	**1.1**
		V	1.7	-	1.4	-
		A	4.6	-	2.4	-
Reducer GB Output bearing	(NDE)	H	1.2	**4.6**	0.8	**1.0**
		V	0.9	-	0.8	-
		A	5.5	-	3.7	-
Ultrasonic level at reducer input bearing		-	**8 LEDs**	-	**4 LEDs**	-

13.7 POST-MAINTENANCE VIBRATION AND ULTRASONIC READINGS

Table 13.1 and Figure 13.14 represent the comparison of pre- and post-maintenance vibration, acceleration, and ultrasound data. All recorded values were reduced to normal level after replacing gearbox input bearing. It was concluded that the success of maintenance work was executed.

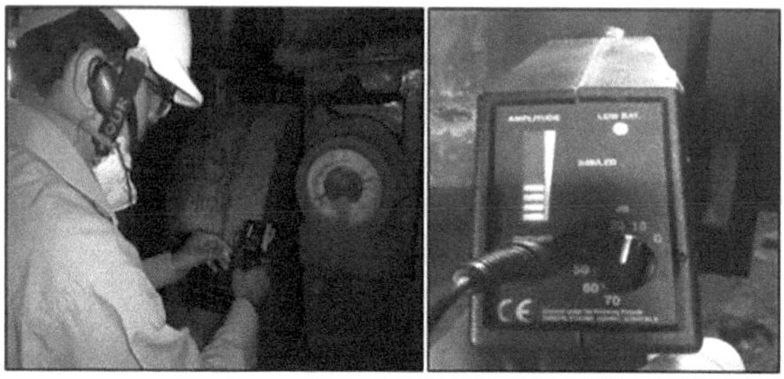

FIGURE 13.14 Post-maintenance ultrasonic level at reducer input DE bearing.

Post-maintenance vibration data and ultrasonic level (depicted in Table 13.1 and Figure 13.14) did not show any abnormality like bearing damage or looseness.

13.8 CONCLUSIONS

Coal conveyor 4A is vital auxiliary which is used to transfer coal from either conveyor 3A/3B directly from rake or from Conveyor 9A/9B (through coal yard) to belt feeder No. 5/6 then subsequently to Conveyor 5A/5B to transport the coal finally at coal bunker level at speed of 2.7 m/s with capacity of 900 TPH (tonnes per hour). It has motor rating of 224 kW with speed of 1485 rpm and at full load, draws 25.8 A current. It is one of the most critical conveyors in coal handling plant, which has the longest inclination. Moreover, it is important for always maintaining healthy bunker levels. Failure of this conveyor will directly affect the operational reliability of bunkering stream and coal supply to boiler will be interrupted, which may lead to power generation loss.

This case study has clearly demonstrated that condition-based vibration testing is a vital component in rotating machinery analysis and detection of abnormality at early stage. Chemical analysis and ultrasonic measurement played a major supportive role to confirm the damage of gearbox input bearing. Acceleration value is a powerful diagnostic information/parameter. Each vibration analyst should use to improve vibration analysis accuracy because this can help maintenance organization to become more proactive and accurate.

The damage of intermediate and output bearings, and gears were avoided using these techniques and other major benefits include such as prevention of downtime, avoidance of major motor repair or replacement costs. The application of coordinated condition monitoring helped in diagnosing the exact root cause of abnormity well in advance.

14 Case Study 13 *Coal Mill 2EF: Gearbox Foundation Deterioration – August 2017*

EXECUTIVE SUMMARY

The present case study is a success story of the implementation of condition monitoring techniques at ADTPS, wherein imminent deterioration of main reducer gearbox foundation of Coal Mill – 2EF was diagnosed. Vibration monitoring and analysis helped in diagnosing the exact root cause of abnormity of reducer gearbox foundation problem at early stage and thus avoided the initiation of secondary internal damages of gearbox. This has saved plant from unplanned outage of critical auxiliary used for pulverizing the coal and associated maintenance cost.

From 13 July 2017 to 15 July 2017, main motor DE bearing was replaced, and pinion backlash was adjusted to address high motor DE axial vibration problem. On 16 July 2017, motor DE axial vibration was reduced from 9.8 to 4.2 mm/s. On 18 July 2017, it was found that horizontal vibration reading deviated from 6.8 to 9.1 mm/s of main reducer input and output bearings. There was no deviation observed in acceleration value.

On 20 July 2017, all process parameters were reviewed, and no other deviation was noticed. On 21 July 2017, readings were taken on reducer gearbox foundation bolts and concrete foundation bolts to pinpoint the problem, and deviation was observed in reducer GB input and output foot vibration reading.

From 22 July 2017 to 26 July 2017, as the vibration level was stable between 9.7 and 10.7 mm/s against the alarm value of 12.0 mm/s, it was decided to keep a close watch on vibration readings. On 27 July 2017, detailed vibration analysis and foundation reading data along with various probable causes were discussed in daily planning meeting.

On 28 July 2017, by getting the high horizontal vibration of main reducer, it was decided to stop coal mill operation for rectification of gearbox foundation. During maintenance, reducer gearbox foundation rectification was carried out by leveling the base and replacing more shims (with less no. of thick shims) and ensuring all machine foots level in one plane. Internal inspection of gears and bearings was carried out and no deviations were noticed.

14.1 PRE-MAINTENANCE HISTORY

During the maintenance period from 13 to 15 July 2017, main motor drive end (DE) bearing was replaced, and pinion backlash was adjusted to resolve high motor DE axial vibration problem (Figure 14.1). On 16 July 2017, coal mill was taken in service and main motor DE axial vibration reading was reduced from 9.8 to 4.2 mm/s (Figure 14.2). On 18 July 2017, it was observed that horizontal vibration reading of

DOI: 10.1201/9781003432616-14

FIGURE 14.1 Main motor drive end (DE) bearing replacement and its alignment.

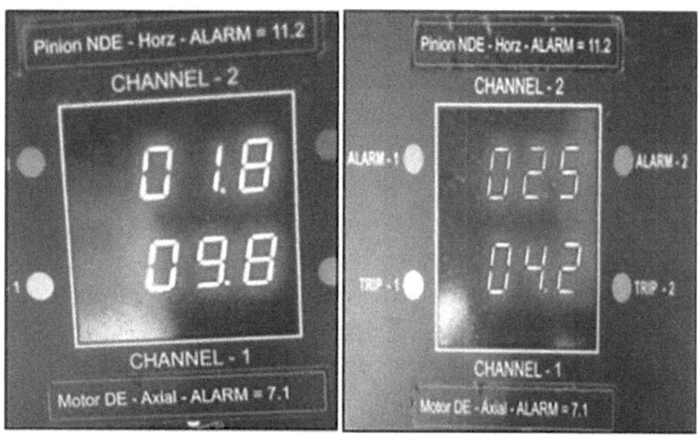

FIGURE 14.2 Pre- and post-online main motor DE axial vibration readings.

main reducer input and output bearing was deviated from 6.8 to 9.1 mm/s (RMS) (Figure 14.3). On 20 July 2017, vibrations of gearbox input and output bearing (horizontal) were found on higher side between 9.7 and 10.7 mm/s at partial and full load respectively (Figure 14.4). No deviation was observed in the acceleration value of gearbox bearing that eliminated the possibility of bearing deterioration (Figure 14.5).

All other process parameters (i.e. temperature, noise) were found normal. On 21 July 2017, for confirming exact root cause of abnormality, readings were taken on main reducer foots and base frame bolts. Deviation was observed in reducer GB input and output foot vibration reading. As the vibration level was stable between 9.7 and 10.7 against the alarm value of 12.0 mm/s, it was decided to keep a close watch on vibration readings. Detailed vibration analysis and foundation reading data along with various probable causes discussed in daily planning meeting. In view of the high horizontal vibration of the main reducer, it was decided to stop coal mill operation for rectification of gearbox foundation.

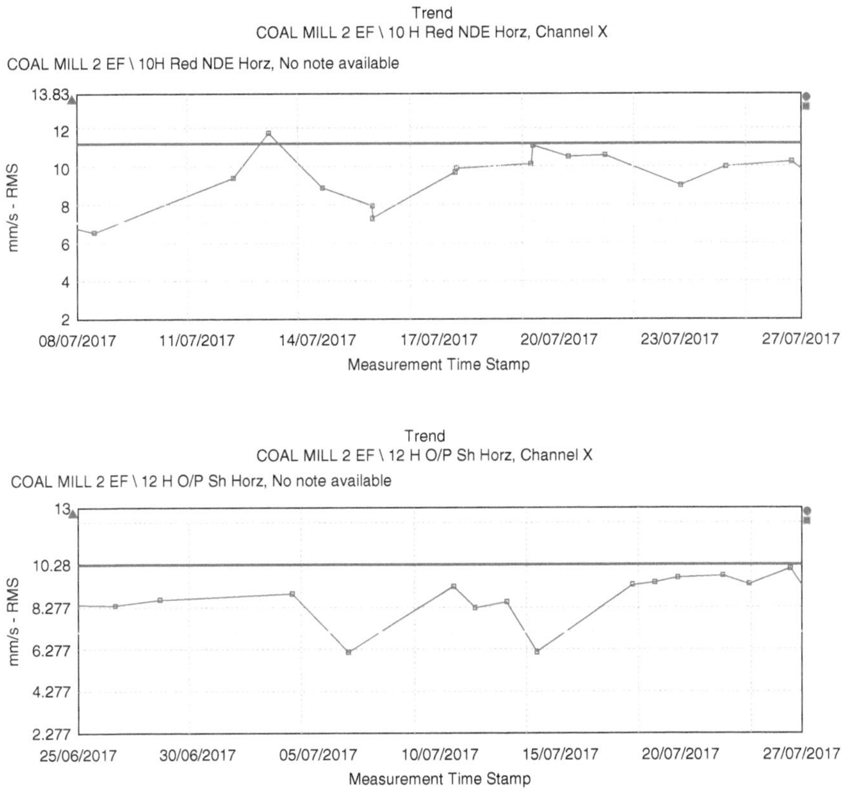

FIGURE 14.3 Vibration trend of gearbox input and output bearing (horizontal).

FIGURE 14.4 Vibration reading of gearbox input and output bearing (horizontal).

14.2 VIBRATION DIAGNOSIS

On 22 July 2017, horizontal vibrations of reducer gear box (input and output) bearing were found to be increased to 10.5 and 10.0 mm/s, respectively. Normally, the same readings remain in the range of 4.5–5.5 mm/s. The spectrum analysis for root cause was investigated as follows:

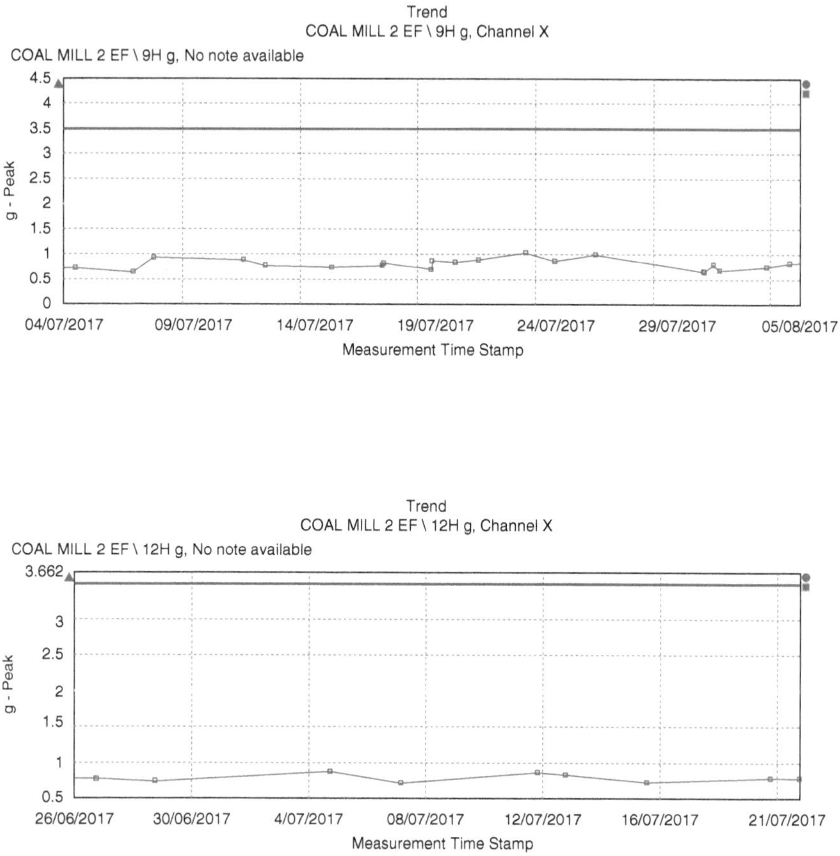

FIGURE 14.5 Acceleration value trend of gearbox input and output bearing (horizontal).

14.3 FREQUENCY ANALYSIS

With reference to Figures 14.6 and 14.7, there was high amplitude observed for horizontal vibration reading of main reducer gearbox (i/p and o/p) bearing and found in between 9.7 and 10.7 mm/s. Spectrum analysis revealed that there was increase in 1× frequency. Acceleration readings of normal bearing eliminated the possibility for existence of any bearing defect (refer to Figure 14.5). Spectrum didn't show gear mesh frequency at 25500 cpm which again eliminated the possibility of existence of any gear defect (refer to Figures 14.6 and 14.7). It was observed that an increment in 1× frequency only in horizontal direction which suspected a foundation problem of the gearbox. To confirm the exact root cause of abnormality, readings were taken on main reducer foots and base frame bolts as shown in Figure 14.8.

With reference to Figure 14.8, after collecting the data, it was observed that base frame foot readings were found in the same range, but main reducer input and output foot readings were not stable and shown that all foots were not in one plane. This

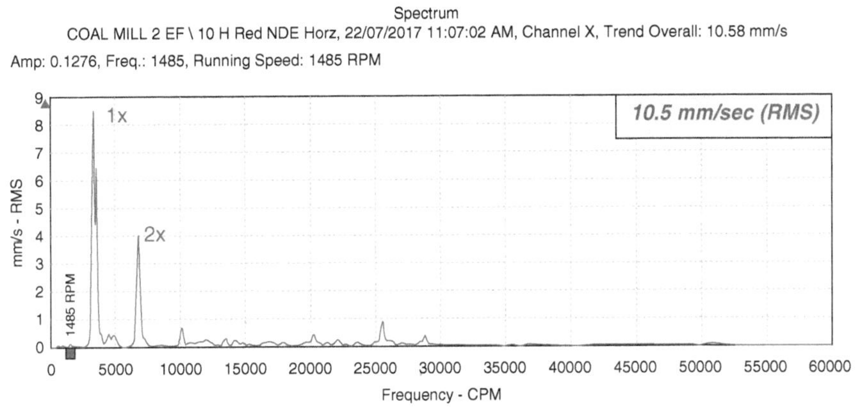

FIGURE 14.6 Vibration spectrum of reducer GB input bearing (horizontal).

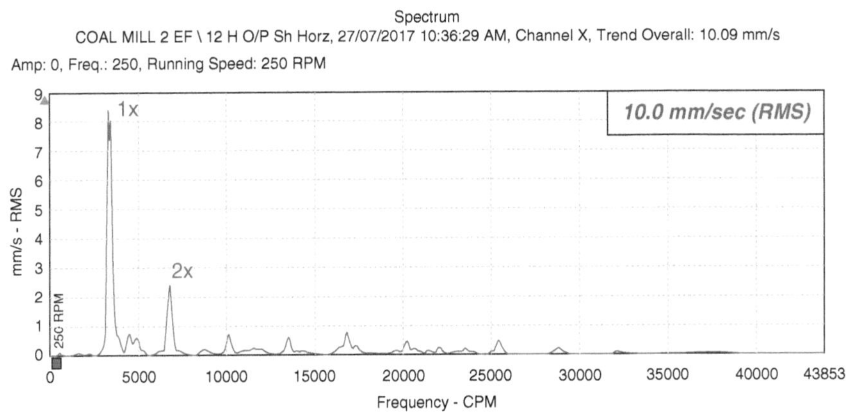

FIGURE 14.7 Vibration spectrum of reducer GB output bearing (horizontal).

confirmed the uneven foot leveling problem of main reducer which was causing high horizontal vibration only to gearbox input and output bearings. The decision was taken to rectify reducer GB foundation and re-align with rest of drive.

14.4 OBSERVATIONS DURING MAINTENANCE WORK

Alignment between motor and reducer GB was checked. All readings were found within limits. Condition of flexible coupling, motor and reducer GB foundation bolts was found good.

14.5 PERFORMED MAINTENANCE WORK

Main reducer gearbox was de-coupled from the rest of group drives. Main reducer foundation foot leveling was done by replacing more shims with less number of thick shims as shown in Figure 14.9.

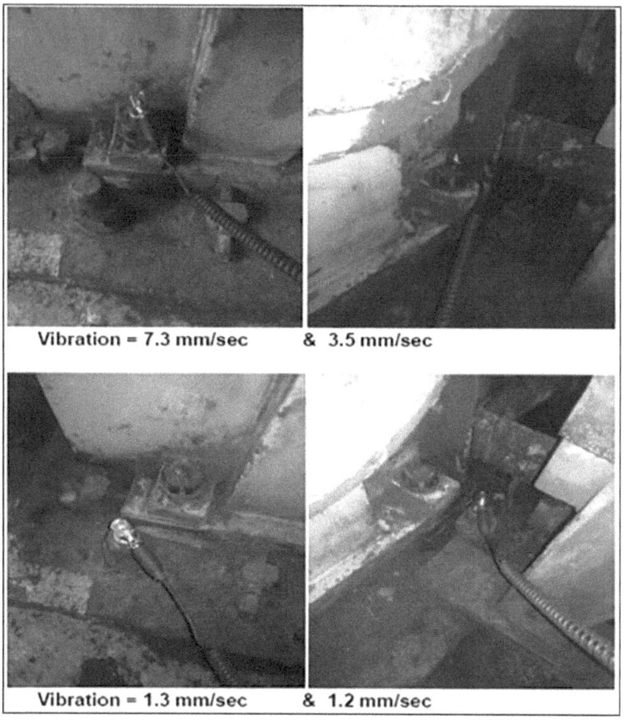

FIGURE 14.8 Vibration data collection at gearbox foot.

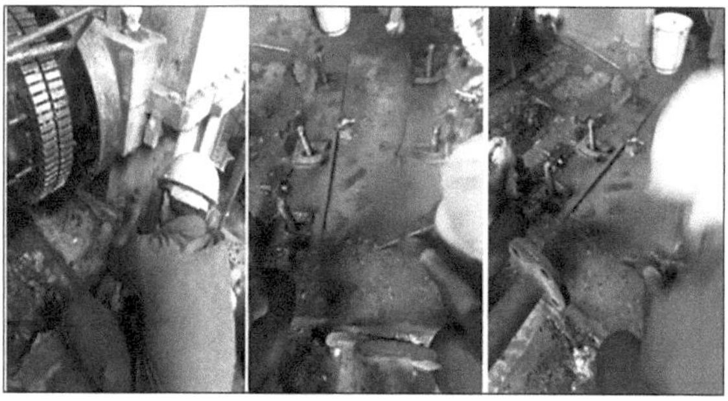

FIGURE 14.9 Gearbox foot foundation leveling done.

Alignment of main reducer GB with rest of drive group was carried out as shown in Figure 14.10. The trial was taken at load condition. Vibrations of motor and reducer GB bearing were found within limit. Coal mill boxed up and made it available for operation.

14.6 POST-MAINTENANCE OFFLINE VIBRATION READINGS

Table 14.1, Figures 14.11 and 14.12 represent comparison vibration data of pre- and post-maintenance. All values were reduced to normal level after repairing main reducer gearbox foundation. It was concluded that maintenance work was executed successfully.

FIGURE 14.10 Final alignment work.

TABLE 14.1
Vibration Report of Coal Mill 2EF

Point		Position	Initial Readings 27/07/2017 Velocity (mm/s) RMS	Initial Readings 27/07/2017 Acceleration (g) Peak	After Foot Leveling of Reducer Gearbox Foundation 01/08/2017 Velocity (mm/s) RMS	After Foot Leveling of Reducer Gearbox Foundation 01/08/2017 Acceleration (g) Peak
Main motor	(NDE)	H	1.4	0.1	0.8	0.1
		V	1.8	-	1.0	-
		A	2.2	-	1.5	-
Main motor	(DE)	H	2.8	0.2	1.5	0.2
		V	3.6	-	1.8	-
		A	4.5	-	6.3	-
Reducer GB	(DE)	H	**10.2**	0.7	**4.4**	0.8
Input		V	7.4	-	4.8	-
bearing		A	6.4	-	4.8	-
Reducer GB	(NDE)	H	**10.2**	0.8	**4.2**	0.9
Input		V	6.8	-	3.3	-
bearing		A	7.2	-	6.2	-
Reducer GB	(NDE)	H	5.3	1.0	2.7	1.1
Intr. bearing		V	4.1	-	3.5	-
		A	7.2	-	6.6	-
Reducer GB	(NDE)	H	**10.0**	0.9	**3.6**	0.9
output		V	5.6	-	4.5	-
bearing		A	6.7	-	5.4	-
Pinion	(DE)	H	5.4	0.8	5.9	0.7
bearing		V	9.0	-	9.0	-
		A	3.2	-	3.8	-
Pinion	(NDE)	H	3.4	0.5	5.0	0.6
bearing		V	6.8	-	9.3	-
		A	1.9	-	3.2	-

14.7 POST-MAINTENANCE REDUCER GEARBOX FOOT VIBRATION READING

FIGURE 14.11 Post-maintenance gearbox foot vibration data collection.

14.8 POST-MAINTENANCE GEARBOX INPUT AND OUTPUT BEARING VIBRATION READING

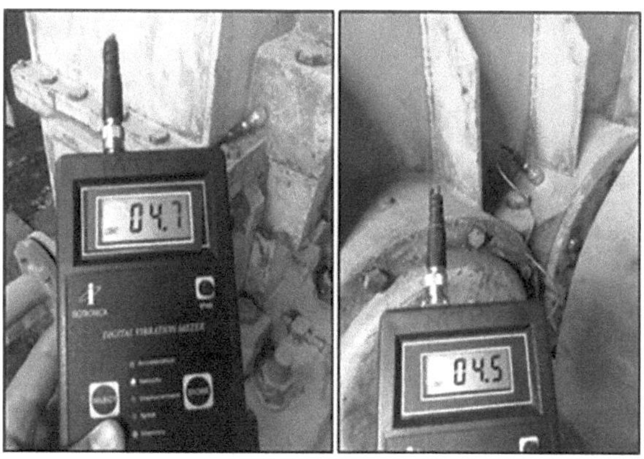

FIGURE 14.12 Post-maintenance gearbox input and output bearing vibration (horizontal).

14.9 CONCLUSIONS

The coal mill is an important auxiliary that is used to crush the coal in the required size to supply in the boiler. Since coal mill supplies coal as fuel to boiler, the reliability needs to be very high. It is an important critical auxiliary in the process. If it fails, it can directly affect the generation loss of 0.125 MUs/h which costs revenue of around Rs.437,500 per hour. There is a total of three nos. of coal mills (i.e. two running at full load and the other is in standby condition). This provision is made because the coal mill is critical equipment in the system and if it fails during the non-availability of stand-by mill, it will lead to tripping of plant and interruption in power supply to our valued customers of Mumbai, India.

The present case study has clearly demonstrated that condition-based vibration testing is a vital component in rotating machinery analysis and detection of abnormality in advance. Vibration monitoring and analysis helped in diagnosing the exact root cause of abnormity of reducer gearbox foundation problem at early stage and thus avoided the initiation of secondary internal damages to gears and bearings. This has saved the plant from unplanned outage of critical auxiliary used for pulverizing of coal and associated maintenance cost. Also, other major benefits include prevention of downtime, avoidance of major gearbox repair or replacement costs.

15 Case Study 14

CW Pump 02: Motor Bearing Looseness – April 2017

EXECUTIVE SUMMARY

Following is a success story of the implementation of condition monitoring techniques at ADTPS, wherein imminent motor bearing looseness of HT auxiliary Cooling Water Pump – 02 was diagnosed.

Vibration monitoring and analysis helped in diagnosing the exact root cause of abnormity at very early stage. This has saved the plant from huge losses in terms of generation, unplanned breakdown time and associated maintenance cost.

On 19 April 2017, during routine monitoring, it was observed that horizontal vibration of motor DE bearing was deviated from 0.59 to 1.89 mm/s (RMS) and 9.7 to 77.5 µm (pk – pk) respectively. On 20 April 2017, proper alignment was done, and vibration reduced from 1.8 to 0.2 mm/s (RMS) and 77.5 to 3.8 µm (pk – pk). On 28 April 2017, again horizontal vibration of motor DE bearing was deviated from 0.2 to 1.9 mm/s (RMS) and 3.8 to 101.5 µm (pk – pk). Vibration of pump DE bearing was stable and found 30 µm. No major deviation was observed. From vibration spectrum analysis, it was suspected looseness of motor bearing. By observing high displacement value, it was decided to take planned outage of the pump for motor replacement. On 29 April 2017, motor was replaced with spare motor and re-alignment was carried out.

15.1 PRE-MAINTENANCE HISTORY

On 19 April 2017, during routine monitoring, it was observed that horizontal vibration of motor drive end (DE) bearing deviated from 0.59 to 1.89 mm/s (RMS) and 9.7 to 77.5 µm peak to peak (pk – pk). As shown in Figure 15.1, on 20 April 2017, alignment was done, and vibration was found to be reduced from 1.8 to 0.2 mm/s (RMS) and 77.5 to 3.8 µm peak to peak (pk – pk). On 28 April 2017, horizontal vibration of motor DE bearing was deviated from 0.2 to 1.9 mm/s (RMS) and 3.8 to 101.5 µm (pk – pk) as shown in Figure 15.2. Also, all motor and pump-related process parameters (i.e. noise, temperature and pressure) were found to be normal. Online and offline vibration of pump DE was stable at 30 µm (pk-pk) and no further deviation was observed.

15.2 VIBRATION DIAGNOSIS

The spectrum analysis for root cause was investigated as follows:

With reference to Figures 15.4 and 15.5, it was found that there was an increase in horizontal vibrations of motor DE bearing. Spectrum analysis revealed that there was an increase in multiples of running speed harmonics from 1×, 2×, 3× up to

 DOI: 10.1201/9781003432616-15

FIGURE 15.1 Alignment correction work.

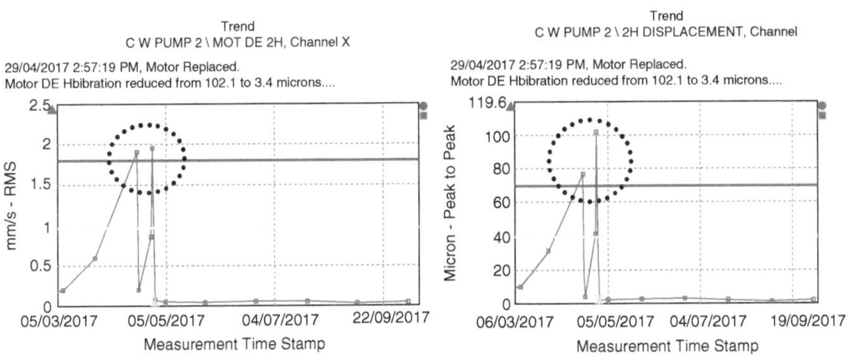

FIGURE 15.2 Vibration trend of motor DE bearing velocity (LHS) and displacement (RHS).

FIGURE 15.3 Online pump bearing vibration reading (microns pk-pk).

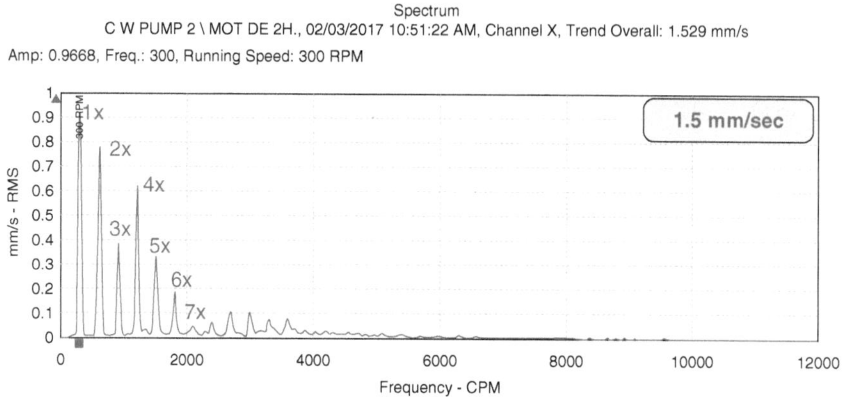

FIGURE 15.4 Vibration velocity spectrum of motor DE bearing - (horizontal).

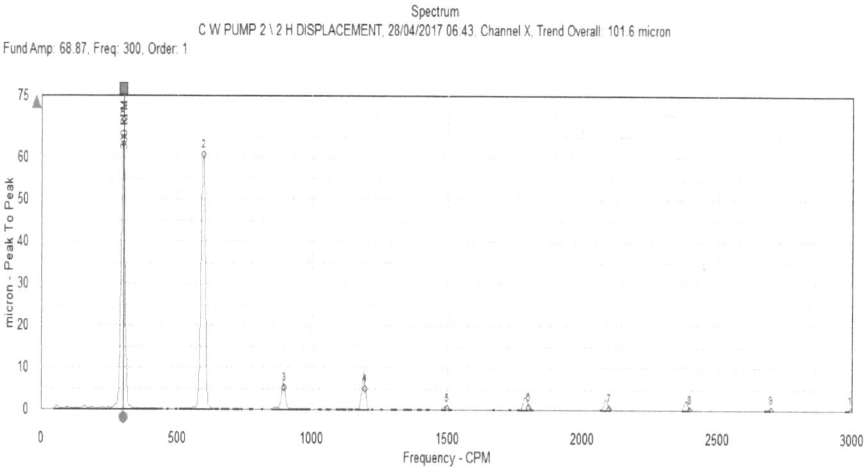

FIGURE 15.5 Vibration displacement spectrum of motor DE bearing - (horizontal).

7× frequency. These results suspected looseness of motor bearing. No deviations were observed in any process parameters of the motor and pump. High displacement value of motor DE bearing (H – 101.5 and V - 44.5 μm – pk-pk) confirmed looseness and misalignment of motor DE bearing. It was decided to stop the pump for motor replacement.

15.3 OBSERVATIONS DURING MAINTENANCE WORK

Gear coupling between motor and pump was checked and found ok as shown in Figure 15.6.

FIGURE 15.6 Inspection of gear coupling between motor and pump.

FIGURE 15.7 Alignment check before motor replacement.

Pump bearing clearances were checked and found ok. Motor bearing and thrust pad clearances were checked and found on the higher side. Alignment was checked from motor nondrive end side and found disturbed by 0.5 mm (see Figure 15.7).

FIGURE 15.8 Motor replaced with spare serviced motor.

15.4 PERFORMED MAINTENANCE WORK

Motor was de-coupled from the rest of group drive. Before dismantling, initial alignment readings were taken for reference. The motor was replaced with a spare serviced one and motor re-aligned with pump as shown in Figure 15.8.

The trial was taken at full load, and vibrations of all motor and pump were found within limit. Colling Water (CW) pump boxed up and made it available for operation.

15.5 POST-MAINTENANCE OFFLINE VIBRATION READINGS

Table 15.1, Figures 15.9 and 15.10 represent a comparison of vibration data for pre- and post-maintenance along with the post-maintenance spectrum. All values were reduced to normal level after replacing motor. It was concluded that maintenance work was executed successfully.

15.6 POST-MAINTENANCE OFFLINE VIBRATION SPECTRUMS

Total Magnitude: 3.418 μm (pk-pk)

Total Magnitude: 1.774 μm (pk-pk)

Post-maintenance vibration spectrums did not show any abnormality such as misalignment or looseness.

TABLE 15.1
Vibration Report of Cooling Water Pump – 02

			Initial Readings		After Replacing Motor	
			28/04/2017		29/04/2017	
			Velocity (mm/s) RMS	Displacement Microns (pk pk)	Velocity (mm/s) RMS	Displacement Microns (pk pk)
Motor	(DE)	H	1.92	101.5	0.07	3.43
		V	0.71	44.5	0.06	1.76
		A	0.23	6.4	0.18	1.64
Pump	(DE)	H	0.73	38.9	0.31	34.3
		V	1.12	50.0	0.35	9.72
		A	0.59	25.4	0.21	7.13
Pump	(DE) Online Reading (μm)	H				

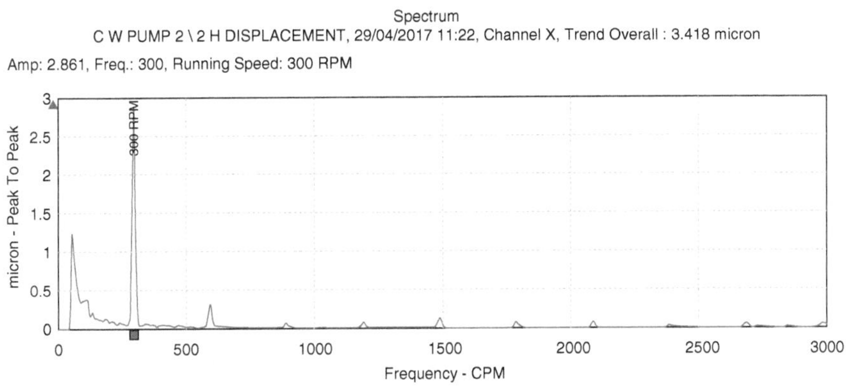

FIGURE 15.9 Vibration spectrum of motor DE bearing (horizontal).

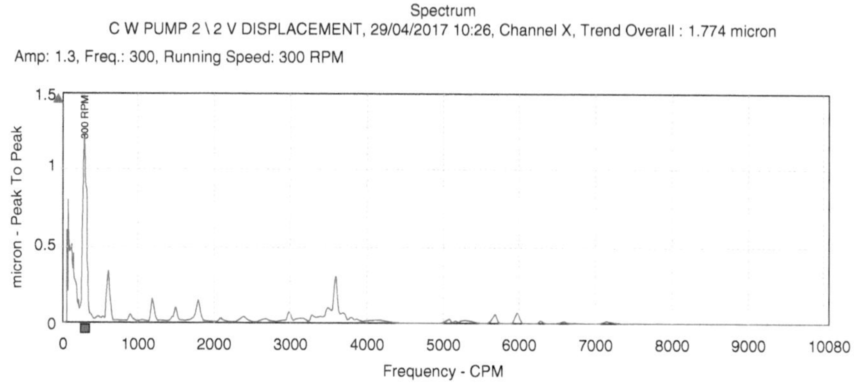

FIGURE 15.10 Vibration spectrum of motor DE bearing (vertical).

15.7 CONCLUSIONS

Cooling water (CW) pump is a critical auxiliary used to supply sea water to condenser for condensation of exhaust steam leaving from steam turbine. Each CW pump is designed to supply sea water of 20750 m^3/h to condenser cooling for the creation of vacuum for safe operation of Low Pressure Turbine (LPT) and is located at 02 km away from condenser. Seawater is also used as Aux Cooling Water (ACW) for cooling of Equipment Cooling Water (ECW) which is used for generator coolers, boiler area via plate-type heat exchangers.

Since this pump supplies CW to the main plant for various purposes, the reliability of pump needs to be very high. There are four pumps of vertical concrete volute-type CW pumps. Each pump has a 50% capacity of total plant requirement. For one unit of 250 MW minimum two nos. of pumps are required on operation mode. During high tide 03 pumps and low tide 04 pumps are required to operate. This provision is made because CW Pump is a critical equipment in the system and if it fails during full load operation, it will tend to deteriorate condenser vacuum that causes generation loss of 0.25 MUS (million units) per hour which may impact financially of Rs. 8,750,00 per hour.

The present case study has clearly demonstrated that condition-based vibration testing is a vital component in rotating machinery analysis and detection of abnormality at early stage. Displacement value played a major supportive role to confirm motor bearing looseness. It is very important to measure for slow-speed machines. The importance of vibration measurement and analysis helped in diagnosing the exact root cause of abnormity well in advance which avoided the damage of motor bearing. Other major benefits include the prevention of downtime, avoidance of major motor repair or replacement costs and reliability of unit.

16 Case Study 15 *Diesel Locomotive 02: Alternator Bearings Deterioration – August 2015*

EXECUTIVE SUMMARY

The present case study is a success story of the implementation of condition monitoring techniques at ADTPS, wherein Alternator bearings deterioration of Locomotive – 02 was diagnosed.

Vibration, noise and ultrasonic analysis of condition monitoring techniques helped in diagnosing the exact root cause of abnormity of alternator knocking sound and deterioration of bearings at engine notch (gear) 3 and 4 speed during load condition at early stage and thus avoided the catastrophic failure of an alternator. This practice has saved the plant from huge losses in terms of demurrage risk and coal availability for electricity generation.

On 27 June 2015, abnormal knocking sound was noticed from alternator at the time of start and stop. Vibration and acceleration reading were taken, and all readings were found well within limits. On 2 August 2015, Diesel locomotive pilot observed that abnormal continuous knocking sound came from alternator side after 4 h of continuous operation at engine notch (gear) 03 and 04 speed. On 4 August 2015, decoupled trial of diesel engine was taken, and no knocking sound was observed which confirmed that the source of the knocking sound was from alternator only.

On 10 August 2015, the acceleration value of alternator DE bearing fluctuates from 0.3 to 1.3 g and NDE from 0.4 to 0.9 g, noise level fluctuates from 101 to 110 Db. Ultrasonic value was also deviated from 5 to 9 LED lights. This shows severity indicator screen of ultrasonic data collector instrument. On 11 August 2015, detailed vibration analysis, noise and ultrasonic monitoring report along with various probable causes were discussed in daily HOD meeting.

On 12 August 2015, greasing of both alternator bearings was done and then the noise level checked in the presence of HODs and observed that all readings were still on higher side. In view of steep increasing trend of acceleration value, noise, and ultrasonic level of alternator bearings, it was decided to stop diesel locomotive operation for replacement of alternator bearings.

During maintenance, minor scoring marks were observed on alternator NDE and DE bearings and the same were replaced. In addition, some scoring marks were noticed on coupling between diesel engine and alternator. This coupling was also replaced on the advice of our superiors considering the larger interest of equipment and system. Detailed inspection of alternator rotor, stator winding, and exciter was carried out in the presence of OEM expert. No damage/scoring marks were observed.

DOI: 10.1201/9781003432616-16

16.1 PRE-MAINTENANCE HISTORY

On 27 June 2015, abnormal knocking sound was noticed from alternator side at the time of start and stop. Vibration and acceleration reading were taken, and all readings were found well within limits. Noise level was measured for reference which was found 98.5 dB (as shown in Figure 16.1) and abnormal sound at the time of start and stop was found to be constant for the next 45 days. No variation in vibration and acceleration value was noticed. Meanwhile, ultrasonic monitoring was started for alternator DE bearing.

On 2 August 2015, loco pilot was observed that abnormal continuous knocking sound came from alternator after 4 h of continuous operation at notch (speed controller) 03 and 04 speed. On 4 August 2015, decoupled trial of diesel engine was taken, and no knocking sound was observed which confirmed that the source of knocking sound was coming from alternator only. It was decided to keep a close watch on all condition monitoring parameters.

On 10 August 2015, readings were taken after 4 h of operation when the knocking sound was continuously observed at load condition. Acceleration value of alternator DE bearing fluctuated from 0.3 to 1.3 g and for NDE bearing from 0.4 to 0.9 g (see Figure 16.2).

After some time, the noise level was fluctuated from 101 to 110 dB (see Figure 16.3) Ultrasonic value was deviated from 5 to 9 LEDs (see Figure 16.4). All parameters compared with LOCO 1 and deviations were observed for LOCO 2 parameters. On 11 August 2015, detailed vibration analysis, noise and ultrasonic monitoring report along with various probable causes were discussed in HoD meeting (see Figure 16.5).

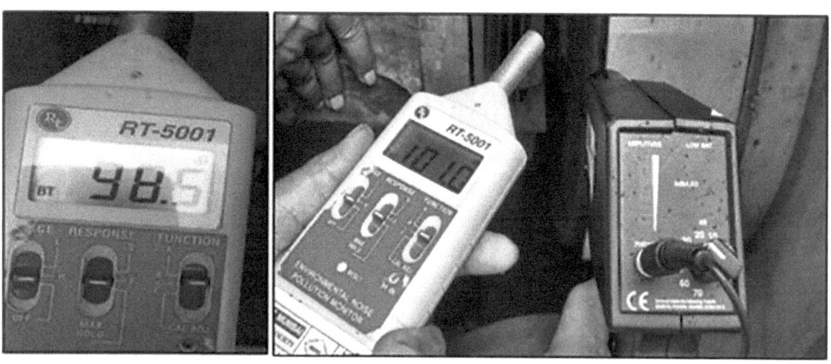

FIGURE 16.1 Noise and ultrasound data collection at alternator bearing.

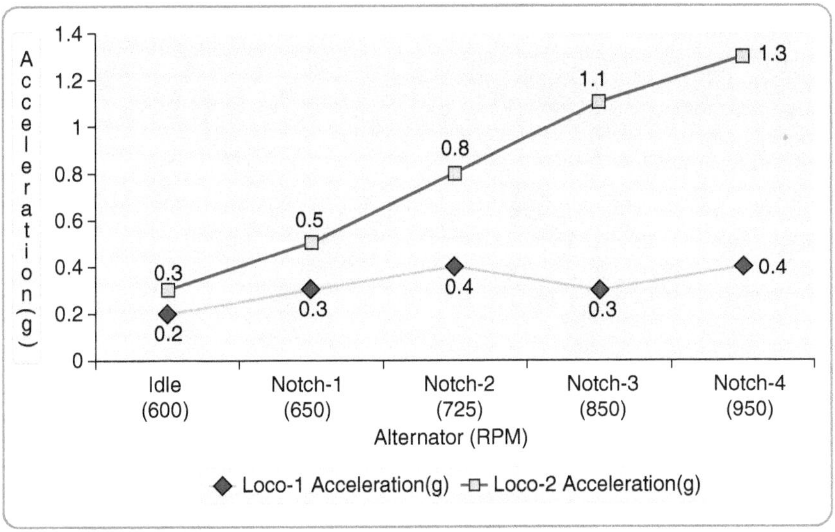

FIGURE 16.2 Acceleration value comparison trend between Loco 1 and 2.

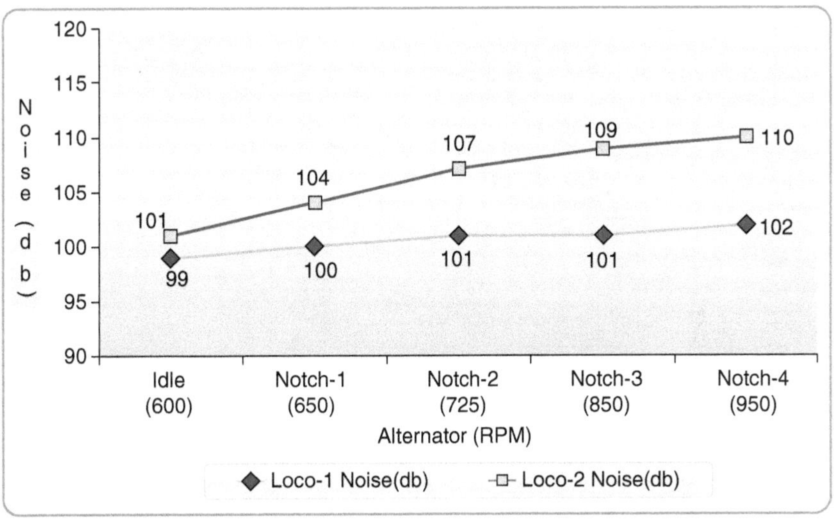

FIGURE 16.3 Noise reading comparison trend between Loco 1 and 2.

Finally, it was decided to grease alternator bearings and collect one more set of readings. Greasing of both alternator bearings was done. On 12 August 2015, the noise level was checked in the presence of HoDs and observed that all readings were still on higher side (Figure 16.6).

On 12 August 2015, in view of steep increasing trend of acceleration value, noise level, ultrasonic level, it was decided to stop LOCO operations for the replacement of alternator bearings.

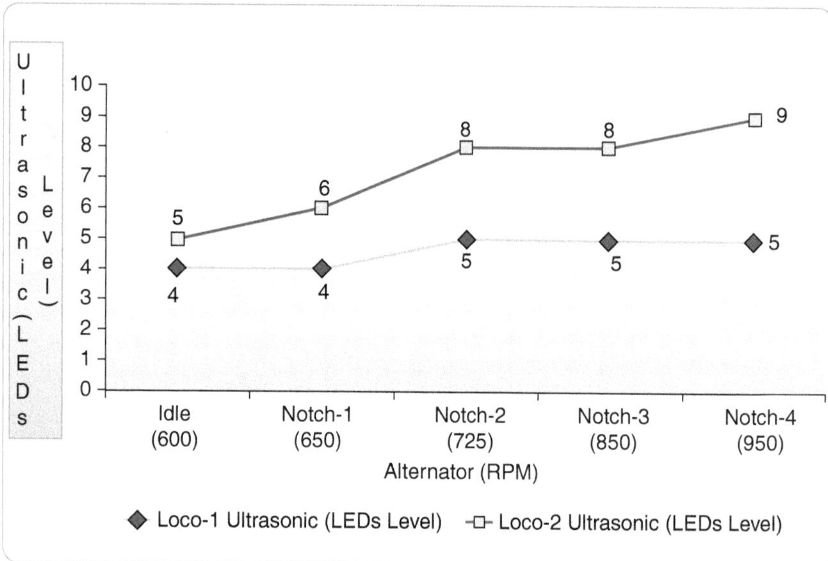

FIGURE 16.4 Ultrasonic value comparison trend between Loco 1 and 2.

Record Notes of Meeting Total ATRs = 0, Pending ATRs = 0, Completed ATRs = 0, DTRs = 0, ITRs = 6

No.	Cat.-Sub Cat.	Minute Narrative	Decision Taken	ID	Priority	Assign To	Pre. Date	Due Date
1	Daily-Defect Review	LOCO-2	Abnormal sound from LOCO Alternator. To be planned tomorrow by CHP & Electrical.	ITR-1	--	--	--	--
2	Daily-Defect Review	Railway- Track	Rail Track up to TG unloading bay inspections to be done by CHP in view of LOCO maintenance planned tomorrow	ITR-2	--	--	--	--
3	Daily-Defect Review	R&D Yard	Overhead line pole foundations crack& deteriorate observed. To be attended on priority by CVP	ITR-3	--	--	--	--
4	Daily-Defect Review	Mechanical Related	Fan lube oil pmp spiders to be replaced periodically to avoid the emergency outages	ITR-4	--	--	--	--
5	Daily-Defect Review	Event analysis report-July-2015	Reviewed.	ITR-5	--	--	--	--
6	Daily-Defect Review	DDR, DPR, DMR, Heat Rate report	Reviewed.	ITR-6	--	--	--	--

FIGURE 16.5 Minutes of plant HoD meeting.

FIGURE 16.6 Greasing for alternator bearings.

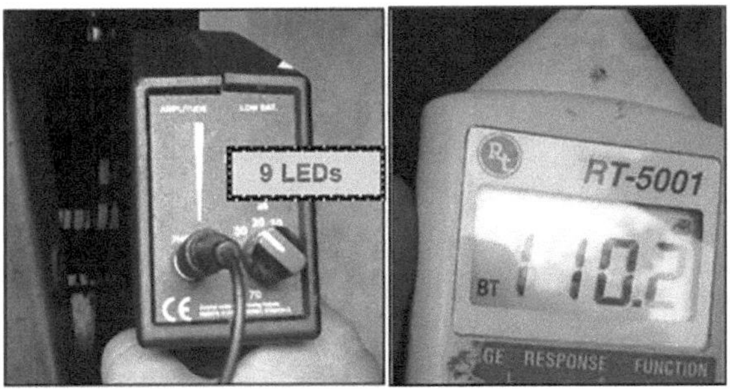

FIGURE 16.7 High ultrasonic and noise reading measured on 27 June 2015.

16.2 ULTRASONIC AND NOISE LEVEL ANALYSIS

Ultrasonic is the application of ultrasound. Ultrasound is used for detection and measurement of high frequencies. Ultrasound is an oscillating sound pressure wave with a high frequency. These high frequencies typically range from 20 to 100 kHz and beyond the range of human hearing. Ultrasound instruments receive high-frequency sounds produced by operating equipment, electrical emissions and leaks. The instruments electronically translate ultrasound frequencies down into the audible range. These frequencies are heard through headphones and observed as intensity on a display panel.

Ultrasonic level was increased from 5 to 9 LEDs as shown in Figure 16.7. This measurement shows ultrasonic reading was on higher side (9 LEDs) which confirmed that alternator bearing deterioration was in progress. In addition, the noise level of alternator was increased from 101.7 to 110.2 dB as shown in Figure 16.7.

16.3 VIBRATION DIAGNOSIS

Non-synchronous value: 0.759; Synchronous value: 0.482; Sub-synchronous value: 0.004.

Non-synchronous value: 1.016; Synchronous value: 0.480; Sub-synchronous value: 0.004.

16.4 FREQUENCY ANALYSIS

With reference to Figures 16.8 and 16.9, acceleration values of Alternator NDE and DE were recorded on higher side (0.9 and 1.3 g respectively). Normally, it was in the range of 0.5 g at load condition. Therefore, it was suspected that motor bearing deterioration was in progress. Acceleration spectrum analysis revealed that there was increase in higher frequencies and the contribution of non-synchronous energy (any vibration that is greater than the 1 × frequency but is not a whole number multiple of it) was at higher side which was mainly due to bearing defects. These high-frequency stress waves' events occur at repeatable time intervals corresponding to a bearing defect frequency, then it was proved that a bearing defect was present. Stress waves are high-frequency vibrations, caused due to impact of two metal surfaces with each other; passing of bearing rolling element over any defect, fatigue cracking, excessive friction, and abrasive wear. Stress waves are short in duration; microseconds for fatigue cracking; milliseconds for metal-to-metal impact.

It was concluded that the acceleration value of alternator bearings was increased due to early-stage deterioration of alternator bearings which in turn increased the noise, ultrasonic level of alternator at higher rpm. It was decided to stop LOCO operations for the replacement of alternator bearings.

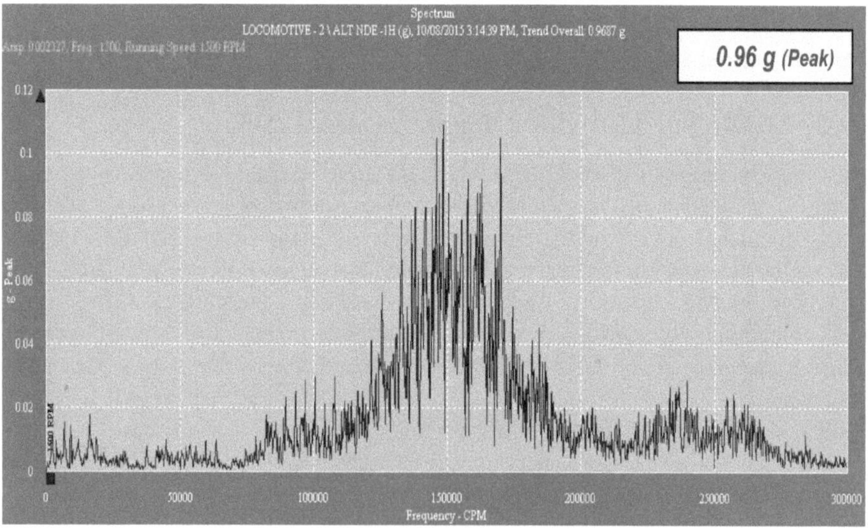

FIGURE 16.8 Acceleration spectrum of alternator NDE bearing (horizontal).

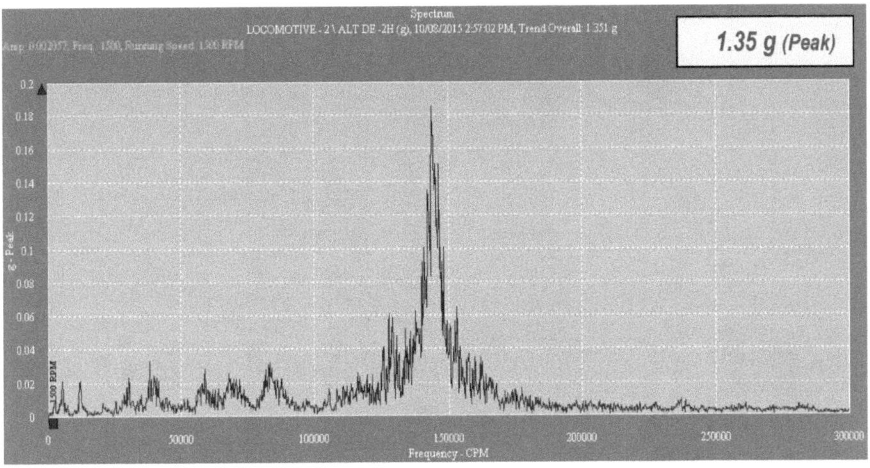

FIGURE 16.9 Acceleration spectrum of alternator DE bearing (horizontal).

16.5 OBSERVATIONS RECORDED DURING MAINTENANCE WORK

Condition of alternator pulley, bearing housing and cooling fan was found okay (see Figure 16.10). Minor scoring marks were observed on both bearings inner race and rollers as shown in Figure 16.11. Also, some scoring marks were observed on coupling between diesel engine and alternator (see Figure 16.12). Alternator shaft surface clearances were checked and noted for reference as shown in Figure 16.13. Detailed inspection of alternator stator winding, rotor and exciter was carried out in the presence of OEM expert (Indian Railways). No scoring marks were observed (see Figure 16.14).

16.6 MAINTENANCE WORK PERFORMED

Initially, alternator sound hood panel was removed as shown in Figure 16.15. Alternator de-coupled from the rest of group drives. Before dismantling, alignment readings were taken for reference (see Figure 16.16). Alternator was taken out from Locomotive and pulley was removed from shaft (refer to Figure 16.17). Alternator bearings was removed and replaced with new bearing (see Figure 16.18) Meanwhile, cleaning and overhauling of all critical associated parts were carried out. Finally, alternator pulley re-mounted on the shaft. By observing scoring marks on coupling (refer to Figure 16.12), it was replaced as per the advice of our superiors considering the larger interest of equipment and system. Assembled alternator was put back at its position (Figure 16.19).

FIGURE 16.10 Proactive check for alternator pulley, bearing housing and cooling fan.

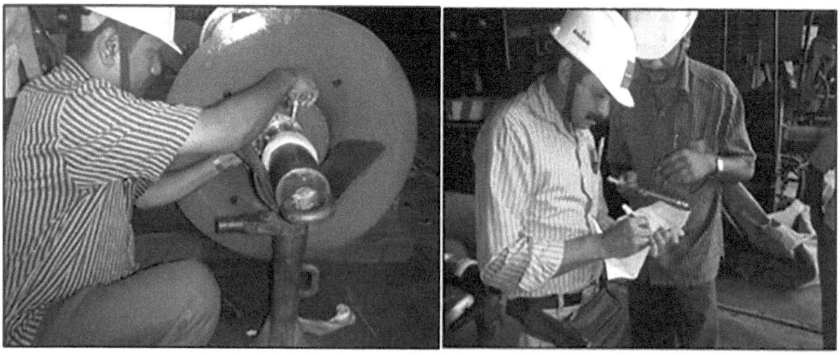

FIGURE 16.13 Shaft dimension check.

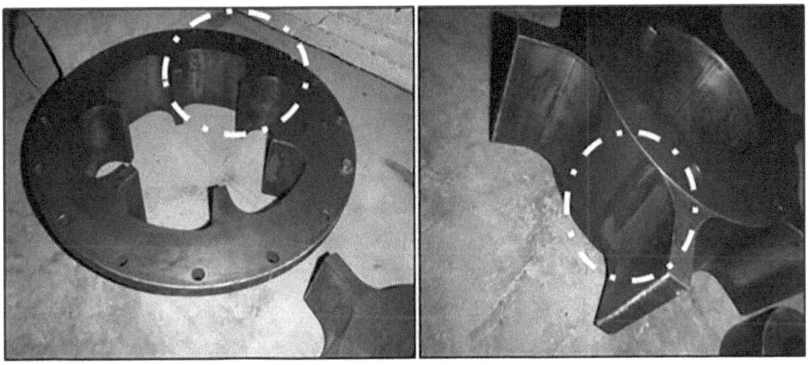

FIGURE 16.12 Minor scoring marks observed on coupling.

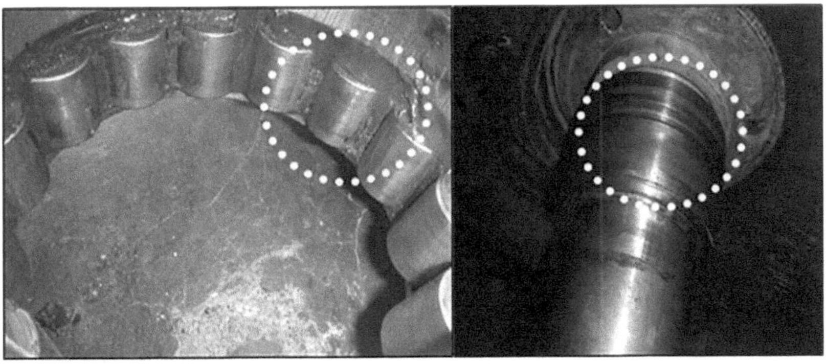

FIGURE 16.11 Scoring marks on alternator bearing rollers.

Alternator was aligned with respect to diesel engine and exciter. The trial was taken, and no abnormal knocking sound was observed from the alternator side at the time of start and stop. Locomotive boxed up and made available for operation. Full load trial (with 16 loaded wagons) was taken, and all readings were found well within limits (Figure 16.20).

16.7 POST-MAINTENANCE OFFLINE VIBRATION READINGS

Table 16.1 and Figures 16.21–16.23 represent a comparison of vibration, acceleration, noise and ultrasound data for pre- and post-maintenance along with their spectrums. All values were found to be reduced to normal level after replacing alternator non drive end (NDE) and drive end (DE) bearings. It was concluded that maintenance work was executed successfully.

FIGURE 16.14 Alternator stator winding, rotor, and exciter inspection by OEM expert.

FIGURE 16.15 Removal of locomotive sound hood panel.

FIGURE 16.16 De-coupling of alternator.

FIGURE 16.17 Removal of alternator for bearing replacement.

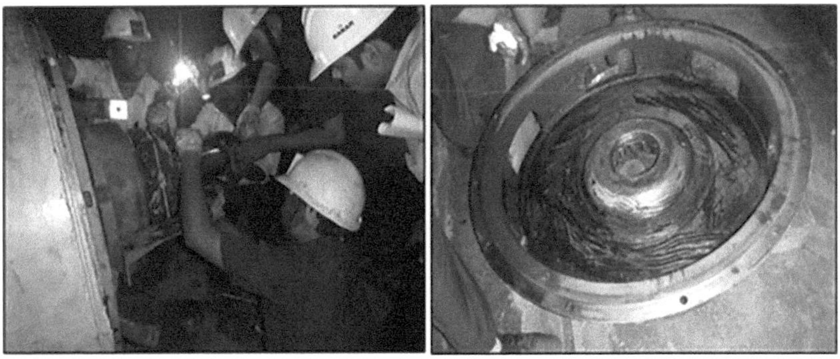

FIGURE 16.18 Alternator bearing replacement.

FIGURE 16.19 Installation of alternator at Locomotive after bearing replacement.

FIGURE 16.20 Locomotive trial at load condition.

TABLE 16.1
Vibration Report of Diesel Locomotive – 02

			At Load and Notch 04 Speed			
			Initial Readings		After Replacing Alternator Bearings	
			12/08/2015		17/08/2015	
Point		Position	Velocity (mm/s) RMS	Acceleration Value (g)	Velocity (mm/s) RMS	Acceleration Value (g)
Alternator	(NDE)	H	2.3	**0.96**	1.6	**0.32**
		V	2.1	-	1.5	-
		A	1.9	-	1.6	-
Alternator	(DE)	H	2.5	**1.35**	1.9	**0.39**
		V	2.6	-	2.0	-
		A	2.1	-	1.7	-
Diesel engine	(DE)	H	3.0	0.8	2.0	0.7
		V	2.1		1.9	
		A	2.3		1.3	
Ultrasonic level		LEDs	**09**		**05**	
Noise level		dB	**110.2**		**102.6**	

16.8 POST-MAINTENANCE NOISE AND ULTRASONIC READINGS

This measurement was depicted that the ultrasonic level was reduced from 9 to 5 LEDs and the noise level also reduced from 110.6 to 102.6dB.

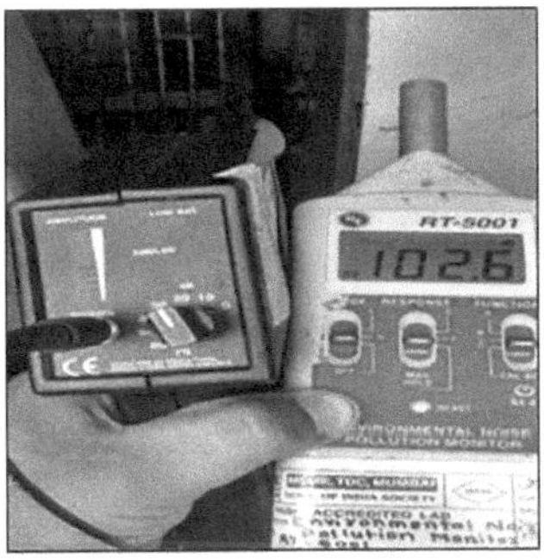

FIGURE 16.21 Post-maintenance ultrasonic and noise data.

16.9 POST-MAINTENANCE ACCELERATION SPECTRUM

Post-maintenance acceleration spectrums did not show any abnormality such as bearing defect frequencies after replacement of alternator bearings.

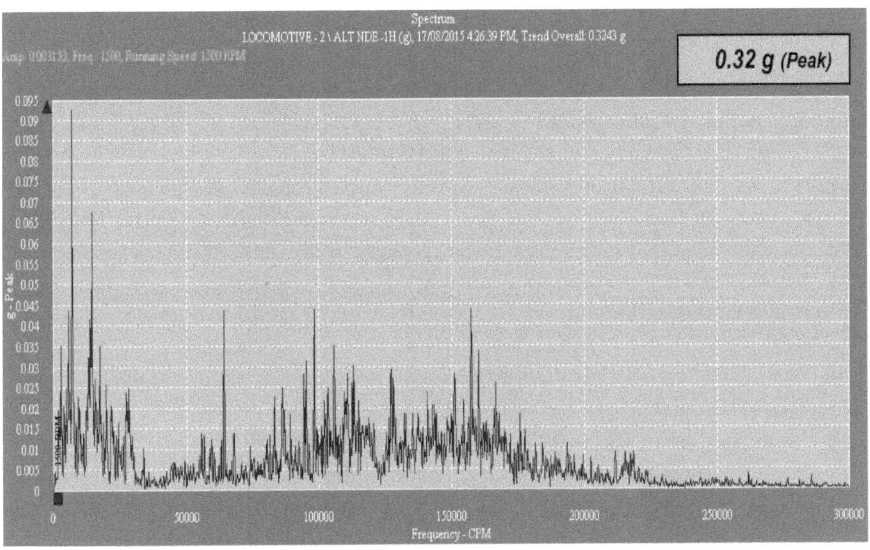

FIGURE 16.22 Reduction in alternator NDE acceleration value from 0.96 to 0.32 g.

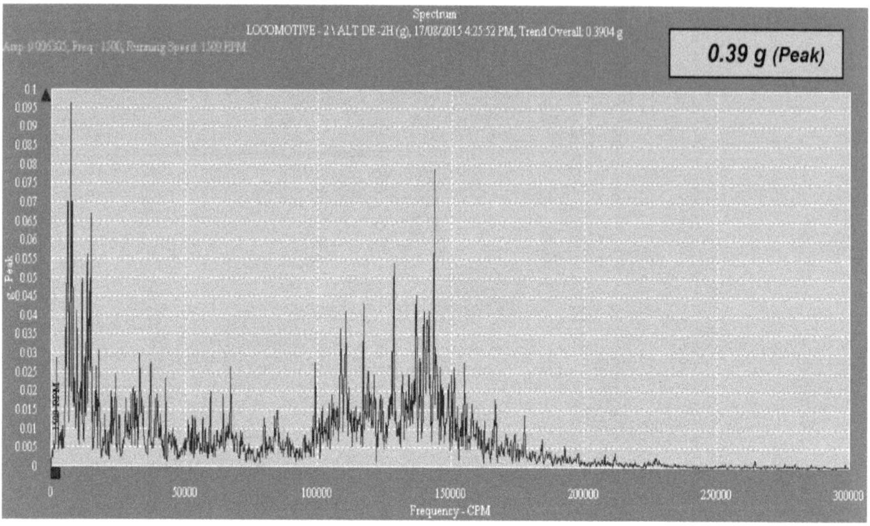

FIGURE 16.23 Reduction in alternator DE Acceleration value from 1.35 to 0.39 g.

16.10 CONCLUSIONS

Co-ordinated condition monitoring techniques such as vibration, noise and ultrasonic analysis helped in diagnosing the exact root cause of abnormity of alternator knocking sound and deterioration of bearings at notch (speed controller) 3 and 4 during load condition at early stage. Thus, catastrophic failure of alternator was avoided.

Presently, ADTPS has 2 nos. of diesel locomotives for hauling the coal wagons to wagon tippler for unloading. These locomotives have been in service for more than 21 years. Considering the coal unloading requirement of ADTPS, the availability of these locomotives cannot be compromised.

Non-availability of locomotive will greatly affect the following:

- Unloading of coal wagons within the stipulated time of 07 h allotted by Railways,
- Heavy demurrage risk (penalty) Rs.150 per wagon per hour,
- Financial loss of Rs. 9000 per hour for 60 wagons,
- Hamper the coal availability for electricity generation,
- In extreme case tripping of plant and interruption in power supply to our valued customers in Mumbai, India.

The present case study has clearly demonstrated that acceleration is a valuable tool in rotating machinery analysis and is especially useful in detecting early and advanced bearing degradation even vibration levels are on lower side. Noise measurement and ultrasonic analysis played a major supportive role to confirm the source of abnormal knocking sound from an alternator. It is a powerful diagnostic parameter. Each vibration analyst should use to improve vibration analysis accuracy as ultrasonic analysis can help maintenance to become more proactive and accurate. The catastrophic

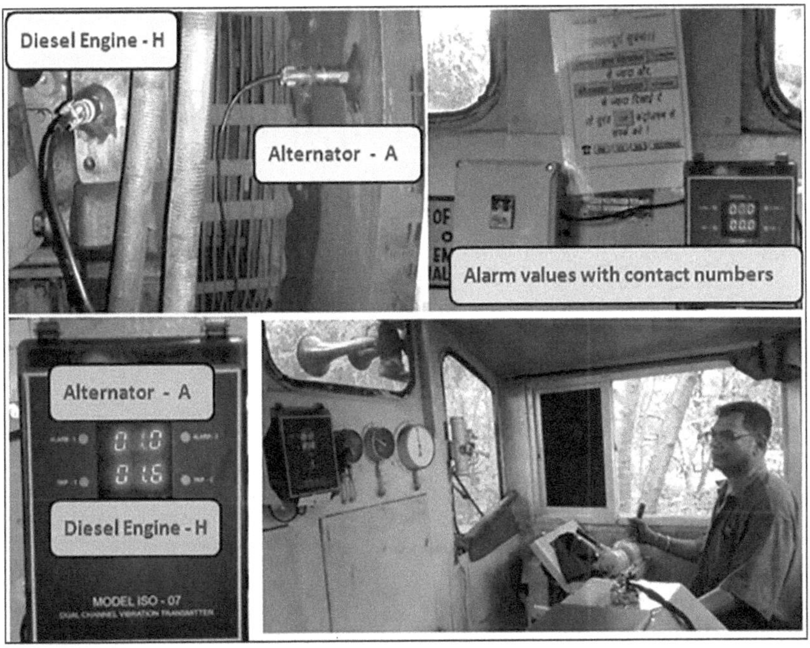

FIGURE 16.24 Online vibration monitoring system for diesel locomotive.

failure of alternator is avoided using these techniques and other major benefit includes prevention of downtime, avoidance of major alternator repair or replacement costs.

With reference to Figure 16.24, to avoid the recurrence of such event and detect the abnormality at early stage, online vibration monitoring system was initiated for both diesel locomotives by installing one sensor for diesel engine drive end (DE) bearing in the horizontal direction and another for alternator drive end (DE) bearing in the axial direction. The online data provided to loco pilot panel with display of alarm and limit values along with the contact number to be used in case of any deviation noticed.

17 Case Study 16

Stacker Reclaimer 01 (CHP): Motor Foundation Deterioration – June 2015

EXECUTIVE SUMMARY

The present case study is a success story of the implementation of condition monitoring techniques at ADTPS, wherein imminent deterioration of bucket wheel drive (used for lifting coal from coal yard) motor foundation of Stacker Reclaimer -01 was diagnosed.

Vibration monitoring and analysis helped in diagnosing the exact root cause of abnormity of motor foundation problem at very early stage and thus avoided the catastrophic failure. This has saved the plant from unplanned breakdown of critical auxiliary used for coal blending process. On 25 May 2015, during routine monitoring, it was observed that vertical vibration of motor DE and NDE bearing were deviated from 7.7 to 14.9 and 7.2 to 14.8 mm/s (RMS) respectively even though vibration of reducer gearbox was normal. It was decided to keep a close watch on vibration readings of motor and reducer gearbox during operation.

On 6 June 2015, vertical vibration of motor NDE and DE was increased to 17.6 and 15.8 mm/s respectively. Vibration spectrum analysis revealed that there is an increase in 1× frequency. On observing the increasing trend of vertical vibration, decision was taken to carry out the necessary action. Increase in 1× frequency only in vertical direction of motor can be due to misalignment between motor and gearbox or foundation problem of the motor. To confirm exact root cause of abnormality, readings were taken on motor and GB foots which showed uneven readings on motor foot which confirmed the foundation problem of motor.

By investigating this problem, motor foundation rectification was carried out by replacing structure pivot pin and providing additional support to motor foundation ensuring all foots are in one plane.

17.1 PRE-MAINTENANCE HISTORY

On 25 May 2015, during routine monitoring, it was observed that vertical vibrations of stacker reclaimer bucket wheel drive motor DE and NDE bearing were deviated from 7.7 to 14.9 and 7.2 to 14.8 mm/s (RMS) respectively. However, the reducer gearbox vibration was found normal (Refer Figures 17.1 and 17.2).

All process parameters (i.e., temperature, noise) were found normal. It was decided to keep close watch on vibration readings of motor and gearbox during operation. On 2 June 2015, vertical vibrations reading of motor NDE and DE were increased to 17.6 and 15.8 mm/s respectively. Observing such an increasing trend of vertical vibration, it was decided to carry out necessary action.

DOI: 10.1201/9781003432616-17

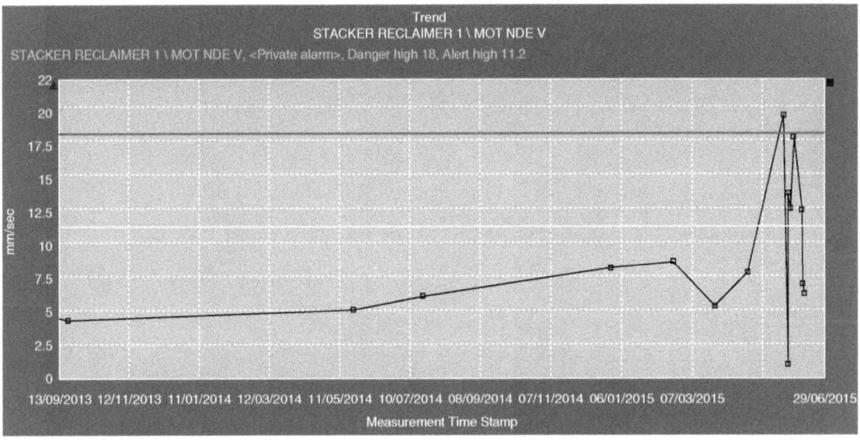

FIGURE 17.1 Vibration trend of motor non-drive end (NDE) vertical.

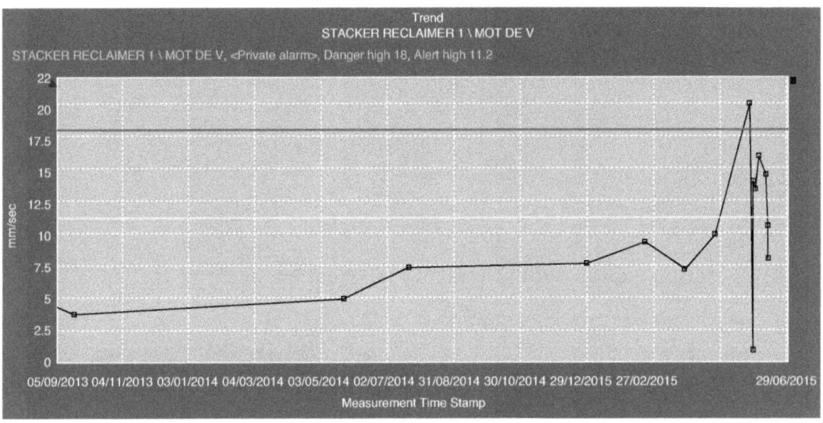

FIGURE 17.2 Vibration trend of motor drive end (DE) vertical.

17.2 VIBRATION ANALYSIS

On 2 June 2015, vertical vibrations of bucket wheel drive motor DE and NDE bearing were increased to 17.6 and 15.8 mm/s respectively (refer Figures 17.3 and 17.4).

The same readings remain normally in the range of 6.0–7.0 mm/s. The spectrum analysis for root cause was investigated as follows:

17.3 FREQUENCY ANALYSIS

There was a gradual increase in the amplitude of vertical vibration of bucket wheel drive motor DE and NDE bearing from 7.7 to 17.6 mm/s and from 7.2 to 15.8 mm/s. Spectrum analysis revealed that there is an increase in 1× frequency (Refer Figures 17.3 and 17.4). Bearing acceleration readings were taken and found normal

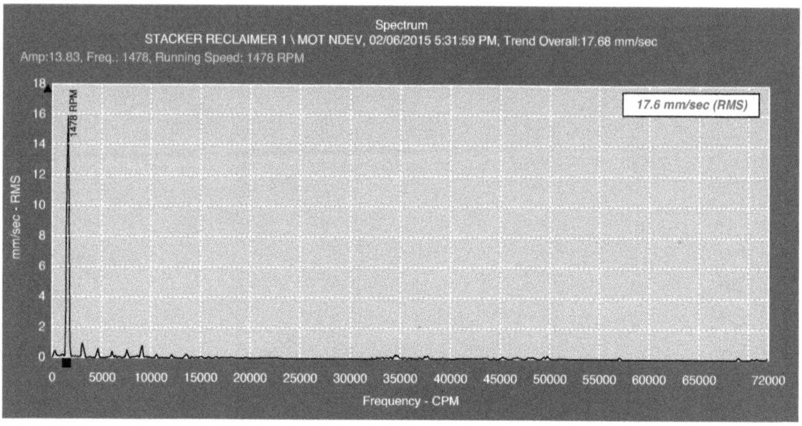

FIGURE 17.3 Vibration spectrum of motor non-drive end (NDE) vertical.

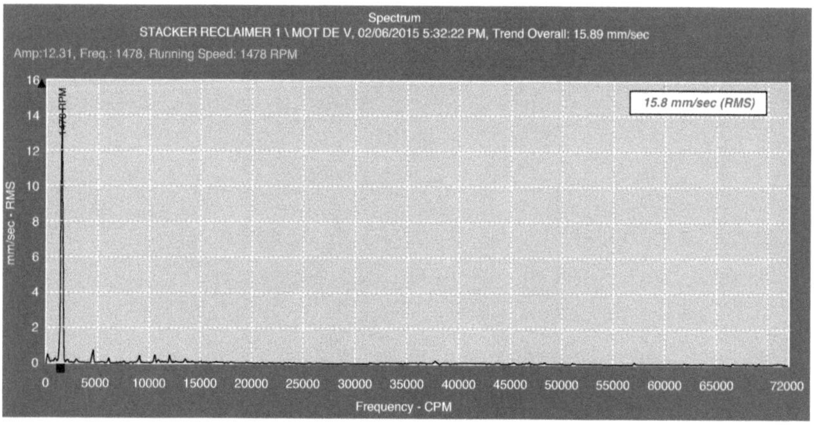

FIGURE 17.4 Vibration spectrum of motor drive end (DE) vertical.

as compared with other stacker bucket wheel motor. While recorded readings were found in the same range with low vibration level which eliminated the possibility of existence of any bearing defect. Increase in 1 × frequency only in vertical direction of motor can be due to misalignment between motor and reducer gearbox or foundation problem of the motor. To confirm exact root cause of abnormality, readings were taken for all motor and reducer GB foots as shown in Figure 17.5.

After collecting the data, it was observed that readings of reducer gearbox foot were found in the same range, but motor foot readings were not stable. This reflected that all four foots of motor were not found in one plane which eliminated the possibility misalignment between motor and reducer gearbox. To confirm the existence of only motor foundation problem, de-coupled motor trial was taken (refer to Figure 17.6).

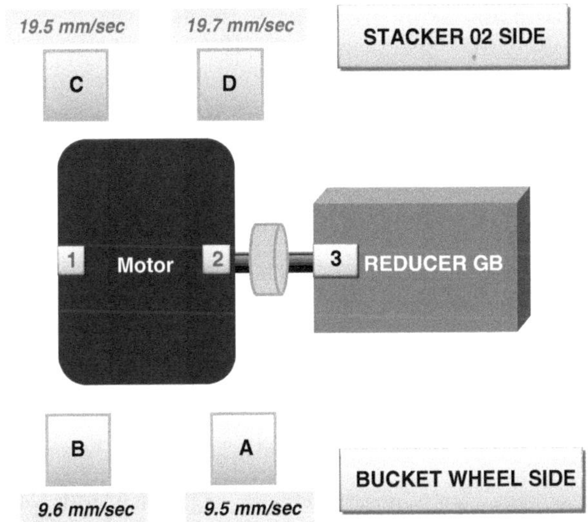

FIGURE 17.5 Motor foot offline vibration data (coupled condition).

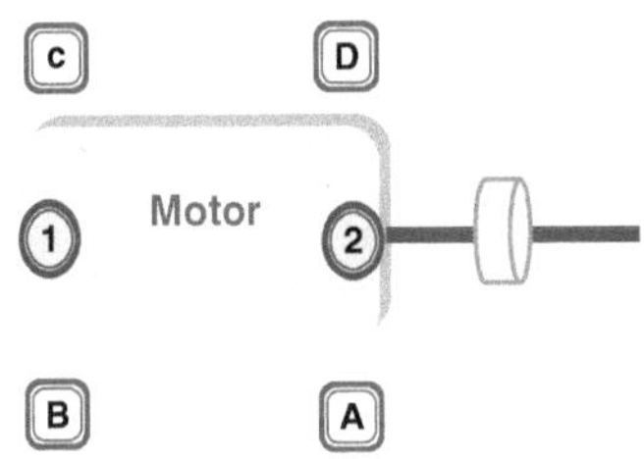

FIGURE 17.6 Motor foot offline vibration data (at decoupled condition).

TABLE 17.1
Motor Foot Offline Vibration Data Comparison

Point Position (as per Figure)	Overall Vibration Velocity (mm/s) RMS
1 – V	0.9
2 – V	1.0
A – V	4.1
B – V	4.3
C – V	9.5
D – V	9.7

From the database of Table 17.1, vertical vibrations of motor DE and NDE bearing were found on higher side with predominant 1× vibration frequency. Also, all motor foot readings showed similar uneven pattern which was found in coupled condition (Figure 17.5). The above analysis confirmed that uneven foundation problem of motor was causing high vertical vibration to the motor and but not to the reducer gearbox. By analysing above vibration data, on 12 June 2015 motor was replaced and again vibration readings were found on higher side with similar pattern at no load de-coupled condition. That's why decision was taken to rectify motor foundation and re-aligned with the rest of the group drives.

17.4 OBSERVATIONS DURING MAINTENANCE WORK

Alignment between motor and reducer GB was checked. All readings were found within limits. Condition of fluid coupling, motor and reducer GB foundation bolts were found well in condition.

17.5 PERFORMED MAINTENANCE WORK

Motor was de-coupled from the rest of group drives. Pivot pin of motor foundation was replaced and foundation strengthening was done by providing additional supports to all motor foots ensuring all foots were remained in one plane (Figure 17.7).

Alignment with the rest of group drives was carried out. Trial was taken at load condition. Vibrations of motor and reducer GB bearing were found within limit. Also, readings were taken on motor foots and they were found in the same range.

FIGURE 17.7 Replacement of motor foundation pivot pin.

Point	Position	24/04/2015	25/05/2015	02/06/2015	09/06/2015	10/06/2015	11/06/2015	21/08/2015
		Initial Normal Readings	First Deviation observed	Gradual increase	Motor Replaced	Pivot pin replaced & Additinoal support provided.		Final Normal readings
		@ load	@ load	@ load	@ no load	@ no load	@ load	@ load
		Velocity (mm/sec) RMS						
Motor (NDE)	H	8.9	9.6	4.4	6.7	4.6	6.1	5.7
	V	7.7	14.0	17.6	12.3	5.9	6.2	4.4
	A	5.7	12.2	6.6	3.2	5.1	2.4	2.1
Motor (DE)	H	9.1	8.4	3.5	7.9	6.1	4.9	4.1
	V	7.2	14.8	15.8	14.4	8.9	8.0	6.7
	A	5.9	9.4	7.4	5.9	4.1	2.9	2.3
Reducer GB P1 (Input)	H	3.2	3.6	3.1	4.9	4.0	3.5	3.6
	V	8.8	11.5	11.6	9.7	7.9	7.6	4.9
	A	3.1	2.3	2.5	1.5	2.8	1.8	1.3
Reducer GB P3 (Output)	H	3.0	2.2	2.4	1.7	3.4	1.6	1.4
	V	6.6	5.9	3.9	4.0	3.1	3.6	2.8
	A	4.3	3.7	3.8	3.7	2.9	2.8	2.6

FIGURE 17.8 Pre- and post-maintenance vibration data.

17.6 POST-MAINTENANCE OFFLINE VIBRATION READINGS

Figure 17.8 and Table 17.2 represent a comparison of vibration data for pre- and post-maintenance of motor bearing and foundation foot at load condition. All values were reduced to normal level after replacing motor foundation pivot pin. It was concluded that maintenance work was executed successfully.

17.7 POST-MAINTENANCE FOUNDATION VIBRATION READINGS

TABLE 17.2
Pre- and Post-Maintenance Motor Foot Offline Vibration Data Comparison

Point Position (as per Fig. 2.16.5)	Initial Readings Velocity (mm/s) RMS (Vertical)	After Rectification in Motor Foundation Velocity (mm/s) RMS (Vertical)
A	9.5	8.1
B	9.6	8.0
C	19.5	12.1
D	19.6	12.5

17.8 CONCLUSIONS

Vibration monitoring and analysis helped in diagnosing the exact root cause of abnormity of motor foundation problem at very early stage. Thus, catastrophic failure of motor was avoided.

Stacker reclaimer is critical equipment in coal handling system, mainly used for stacking and reclaiming coal in respective coal yards. It is a unique machine that can either stack or reclaim coal as per process requirement. Since it is operable along the length of the coal yard, gives tremendous ease in coal yard management.

Non-availability of stacker reclaimer will greatly affect:

- Coal yard management (Stacking and Reclaiming)
- Coal blending of different grade which in turn the desired coal blending/fuel input to boiler.
- Increase dozer operation for yard management, resulting in higher diesel consumption.
- Dozer diesel average consumption will increase by 35 L/h, causing financial loss of Rs. 3000 per hour.

The present case study has clearly demonstrated that condition-based vibration testing is a vital component in rotating machinery analysis and detection of abnormality well in advance. The catastrophic failure of motor is avoided using this technique and other major benefit includes prevention of downtime, avoidance of motor repair or replacement costs.

18 Case Study 17
PA Fan 2B: Motor Foundation Deterioration – November 2014

EXECUTIVE SUMMARY

Following is a success story of the implementation of condition monitoring techniques at ADTPS, wherein imminent Concrete Foundation damage of HT auxiliary PA FAN – 2B was diagnosed.

The present case study has clearly demonstrated that condition-based vibration testing is a vital component in rotating machinery analysis and detection of abnormality at early stage. Ultrasonic pulse velocity test, rebound hammer test and chemical test of concrete acted as a major supportive role in confirmation of damage of motor foundation.

Vibration and civil foundation in Residual Life Assessment (RLA) study analysis helped in diagnosing the exact root cause of abnormity. This has saved the plant from huge losses in terms of generation, unplanned breakdown, downtime and associated maintenance costs which could have caused total cost of approximately Rs.200,000.

During November 2012, during routine monitoring, it was observed that horizontal vibration of motor NDE and DE bearing deviated from 1.9 to 2.3 mm/s and gradually increased to 2.8 mm/s in the next 2 months. Readings gradually increased to 3.3 mm/s in the month of November 2013. An uneven vibration was observed on all motor foots. It was suspected foundation problem of motor from vibration spectrum analysis. Civil foundation RLA study confirmed that motor foundation deterioration was in progress. Decision was taken to repair the motor foundation as per suggestions received from civil engineering team. During maintenance, it was observed that crack on foundation top region (extended slab) and the corrosion of rebars were found to be quite alarming and can fail in the near future.

18.1 PRE-MAINTENANCE HISTORY

From June 2011 to October 2012, horizontal vibration readings of motor NDE were found stable and ranged between 1.6 and 1.9 mm/s (Figure 18.1). From November 2012, minor variation was observed in vibration reading of motor NDE and DE in only horizontal direction from 1.9 to 2.3 mm/s and gradually increment was observed up to 2.8 mm/s for the next 2 months as shown in Figure 18.2. Also, all other process parameters (i.e., airflow, pressure, temperature) were found normal. On 9 January 2013, notification (10000036300) was raised on SAP software to mechanical maintenance department to check the alignment during possible opportunity. High vibration issue along with probable causes was discussed in daily HoD

DOI: 10.1201/9781003432616-18

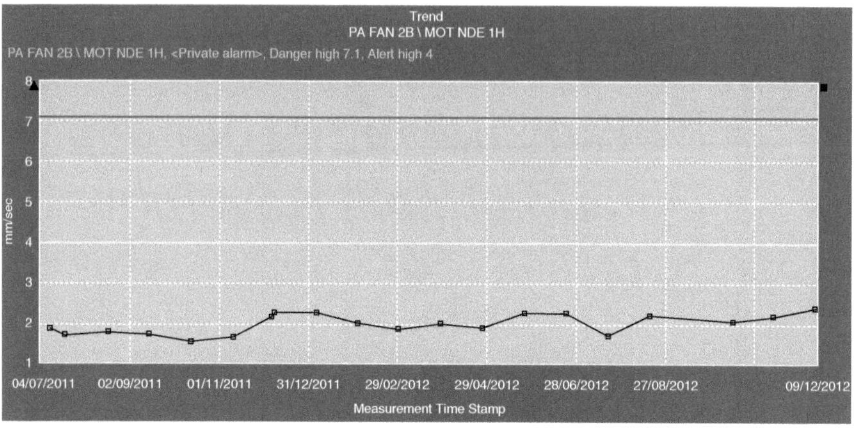

FIGURE 18.1 Vibration trend of motor NDE bearing – horizontal (initial).

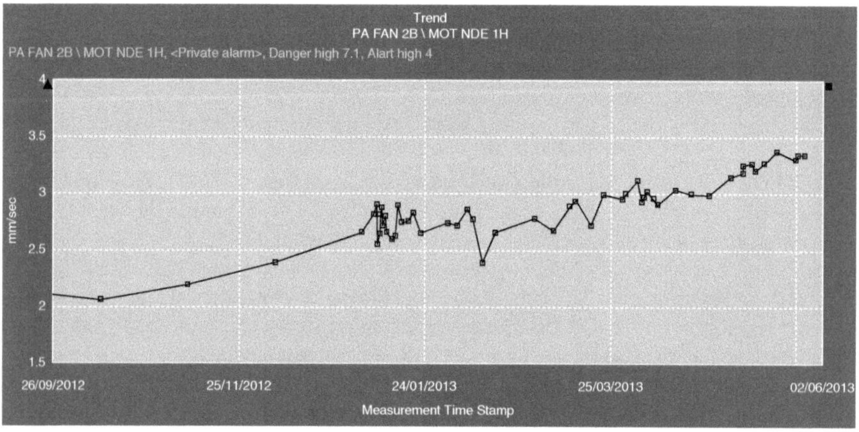

FIGURE 18.2 Vibration trend of motor NDE bearing – horizontal (after 02 months).

meeting. On 18 January 2013, alignment was checked, and minor correction was carried out during short shutdown. The horizontal vibration of motor NDE and DE was reduced from 2.8 to 2.6 mm/s and remained stable for the next 15 days. It was decided to keep a close watch on horizontal vibration readings of motor bearing on a daily basis. From 7 February 2013, again vibration readings increased from 2.6 to 2.9 mm/s (see Figure 18.3) and remained stable for next 01 month. On 3 March 2013, motor vibrations increased from 2.9 to 3.0 mm/s. Meanwhile, readings were taken on all motor foot (see Figure 18.4). Uneven vibrations were observed on all motor foot (shown in Table 18.1). From Figure 18.4 and Table 18.1, uneven foundation problem of motor was suspected which was causing high horizontal vibration to the motor. On 30 March 2013, notification (100000038808) raised on SAP software to civil maintenance department for repairing the motor foundation (Figure 18.5). To confirm the foundation problem, on 5 June 2013, ultrasonic pulse velocity test, rebound hammer test and chemical test of concrete were carried out. As shown in Figures 18.6 and 18.7 and based on the observations, results of various tests were conducted, and

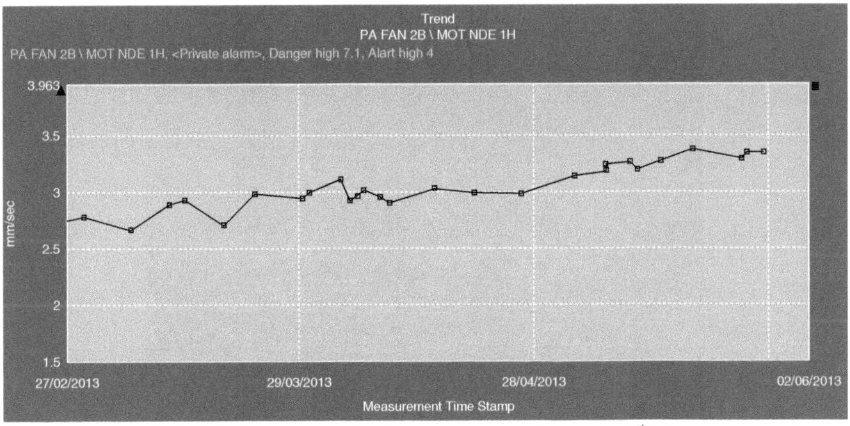

FIGURE 18.3 Vibration trend of motor NDE bearing – horizontal (after alignment).

FIGURE 18.4 Motor foot offline vibration data collection (coupled condition).

TABLE 18.1
Motor Foot Offline Vibration Data Comparison

Point Position (as per Fig. 2.17.4)	Overall Vibration (Vertical) Velocity (mm/s) RMS
01	0.42
02	0.58
03	0.98
04	0.65

experts of civil team had suggested the repairing of damaged foundation by structure strengthening along with provision of additional supports to motor. Observations from Residual Life Assessment (RLA) study were discussed in HoD meeting and an action plan was prepared for the suggested rectification of motor foundation. Repair work was carried out as per suggestions of expert civil team during short shutdown opportunity between 9 and 12 November 2013.

FIGURE 18.5 Abnormalities noticed for motor foundation.

FIGURE 18.6 Ultrasonic pulse velocity test.

FIGURE 18.7 Rebound hammer test.

18.2 VIBRATION DIAGNOSIS

On 8 November 2013, horizontal vibrations of motor NDE and DE bearing reached 3.3 mm/s. The same readings remain normally in the range of 1.2–1.5 mm/s. The spectrum analysis for root cause was investigated as follows:

From Figures 18.8 to 18.9, it was found that there was an increase in only horizontal vibrations for motor NDE and DE bearing. Spectrum analysis revealed an increase in 1× frequency which suspected foundation problem of motor. To confirm foundation damage, ultrasonic pulse velocity and rebound hammer test were carried out.

18.3 ULTRASONIC PULSE VELOCITY TEST AND ANALYSIS

This test is done to assess the quality of concrete as per IS: 13311 (Part 1) –1992. The principle of this test method consists of measuring the travel time of an ultrasonic pulse passing through the concrete being tested. Comparatively higher velocity is obtained when concrete quality is good in terms of density, uniformity, homogeneity. Ultrasonic pulse velocity test was carried out on the foundation concrete by forming grids in all three possible methods (direct, semi-direct and indirect) to get the information about interior concrete.

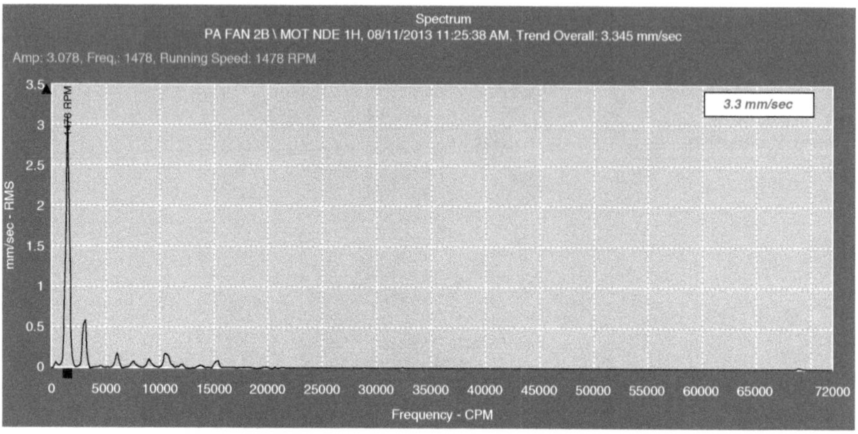

FIGURE 18.8 Spectrum of main motor NDE bearing (horizontal).

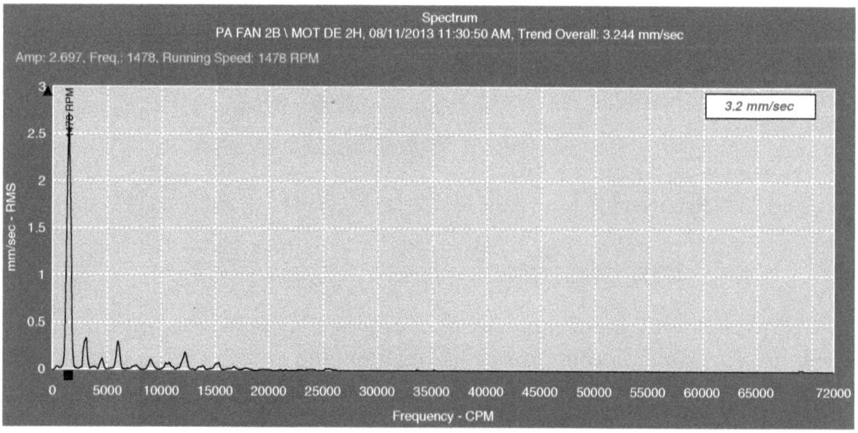

FIGURE 18.9 Spectrum of motor DE bearing (horizontal).

Inferences/results of test conducted were as follows:

- The result indicates that the existing foundation concrete was not dense throughout.
- This could be due to de-bonding of cover concrete at tested locations or the presence of honeycombed interior concrete or a combination of both.
- Hence, suitable treatment for interior concrete needs to be carried out, in addition to other measures.

18.4 REBOUND HAMMER TEST AND ANALYSIS

The rebound hammer test is carried out to find out the compressive strength of concrete by using rebound hammer as per IS: 13311 (Part 2) – 1992. When the plunger of the rebound hammer is pressed against the surface of the concrete, the spring-controlled mass rebounds. Extent of such a rebound depends upon the surface hardness of the concrete.

Rebound hammer test was carried out on the foundation concrete at random in the exposed areas to assess the surface hardness and the strength of concrete near to surface (see Figure 18.10).

Inferences/results of test conducted were as follows:

- The test results indicated that the surface was hard and sound at many tested locations.
- Further, in-situ strength of concrete near to surface was found in the range of 25–36 N/mm^2.
- In some of the other locations, readings could not be obtained due to the debonding of cover concrete.

18.5 CHEMICAL TEST OF CONCRETE SAMPLE

Chemical analysis of concrete can provide extremely useful information regarding the causes of failure of concrete. The tests most frequently carried out for detecting the following parameters:

- Depth of carbonation
- Chloride content
- Cement content
- Sulfate content
- Alkali content

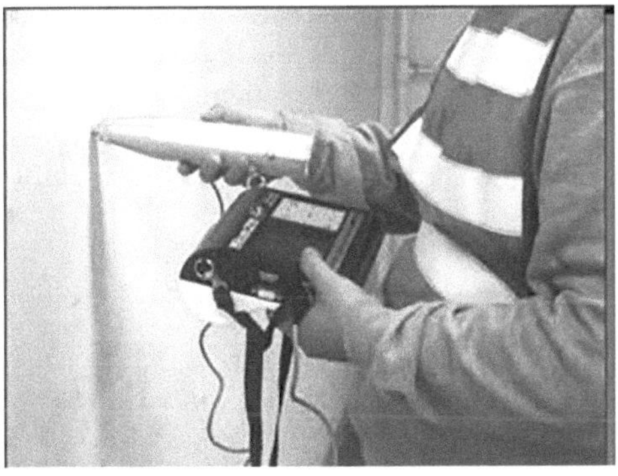

FIGURE 18.10 Rebound hammer test.

FIGURE 18.11 Chemical testing of concrete.

Concrete samples from the spalled region were collected at site for carrying out chemical analysis to find out the level of chlorides, sulfates and pH for the assessment of distress.

Inferences/results of test conducted were as follows:

- The results of chemical test on concrete indicate that the level of chlorides was just on the borderline
- Sulfates and pH were found well within the permissible limits.

18.6 VIBRATION FREQUENCY AND CIVIL RLA STUDY ANALYSIS

There was an increase in amplitude of horizontal vibration of motor NDE and DE bearing from 1.5 to 3.3 mm/s. Earlier, it was found in the range of 1.2–1.5 mm/s. Spectrum analysis revealed that there was an increase in 1× frequency. Readings were taken on all motor foots and observed uneven readings on all foots. It was suspected that possibility of motor foundation may have been damaged. To confirm the foundation problem, ultrasonic pulse velocity test, rebound hammer test and chemical test of concrete were carried out (refer to Figures 18.7, 18.10, and 18.11). RLA study confirmed that deterioration of motor foundation was in progress. Therefore, the decision was taken to repair motor foundation as per suggestions of expert civil team during possible shutdown opportunity.

18.7 OBSERVATIONS DURING MAINTENANCE WORK

Vibrations in the foundation were felt during the operation of equipment (see Figure 18.12).

The observed crack on foundation top region (extended slab) was found to be quite alarming as shown in Figure 18.13.

Spalling of cover concrete on the side face of foundation was observed at a few locations. Exposure of corroded rebars was observed wherever spalling of concrete had taken place (refer to Figure 18.14).

FIGURE 18.12 Motor foundation inspection.

FIGURE 18.13 Crack on motor foundation.

FIGURE 18.14 Exposure of corroded rebars to motor foundation.

18.8 PERFORMED MAINTENANCE WORK

Motor foundation and slab repairing were carried out as per the RLA study suggestions (Figure 18.15).

Additional supports were provided to motor foundation as per the design (see Figure 18.16).

18.9 POST-MAINTENANCE OFFLINE VIBRATION READINGS

Tables 18.2 and 18.3 represent the comparison of vibration data for pre- and post-maintenance of motor bearing and foundation foot at load condition. All values were reduced to normal level after repairing motor foundation and providing additional support. It was concluded that maintenance work was executed successfully.

FIGURE 18.15 Repairing of motor foundation.

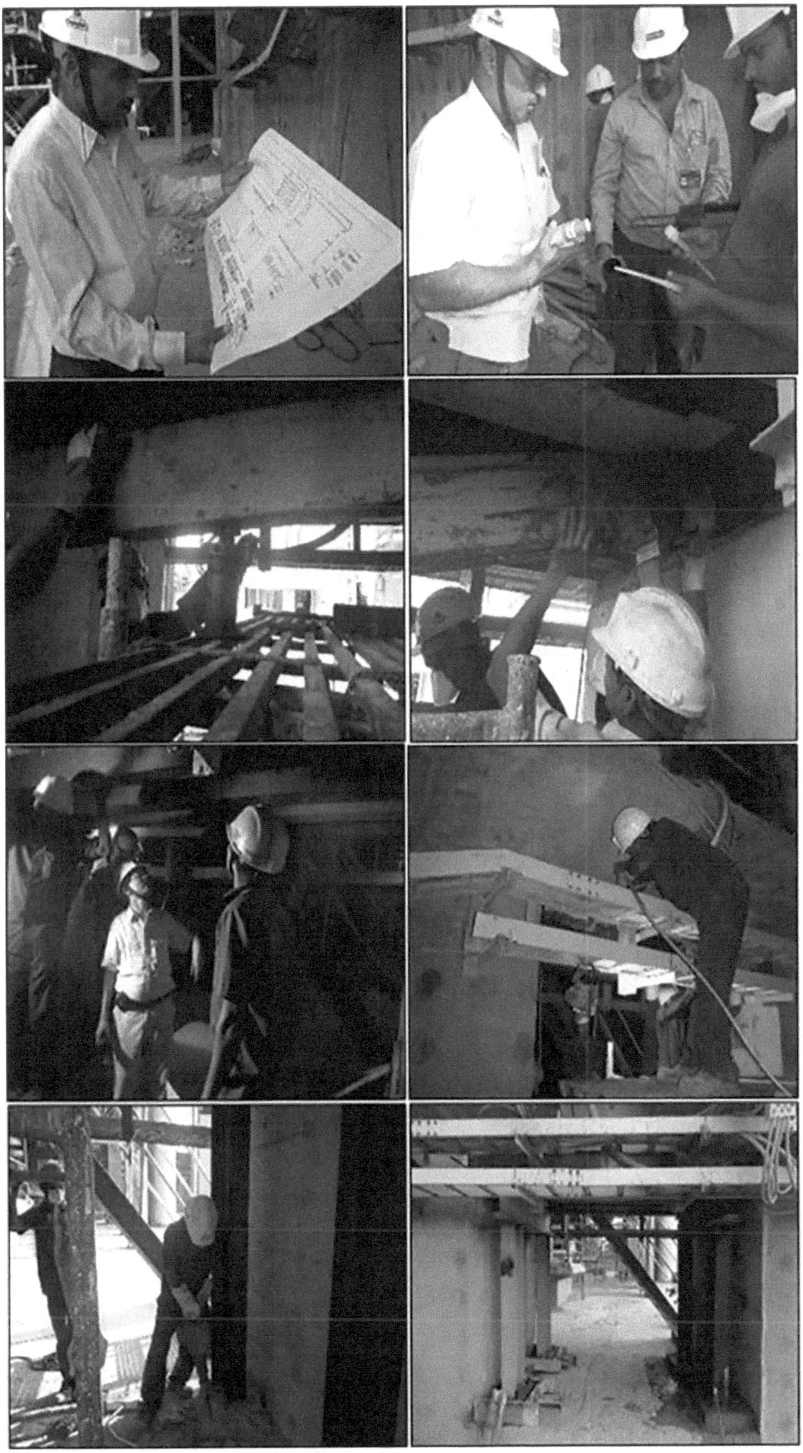

FIGURE 18.16 Provision of additional supports done as per drawing.

TABLE 18.2
Vibration Report of PA Fan 2B

Point		Position	Initial Readings 08/11/2013	After Repairing Motor Foundation 13/11/2014
			Velocity (mm/s) RMS	
Motor	(NDE)	H	**3.3**	**1.3**
		V	0.6	0.6
		A	0.5	0.6
Motor	(DE)	H	**3.2**	**1.4**
		V	0.5	0.6
		A	0.6	0.5

18.10 POST-MAINTENANCE FOUNDATION VIBRATION READINGS

TABLE 18.3
Comparison of Vibration Ddata for Pre- and Post-maintenance Motor Foot (Offline)

Point Position (as per Fig. 2.17.4)	Overall Vibration (Vertical) Velocity (mm/s) RMS	
	Initial Readings	After Repairing Motor foundation
01	0.42	0.31
02	0.58	0.32
03	0.98	0.44
04	0.65	0.35

18.11 POST-MAINTENANCE VIBRATION TREND AND SPECTRUMS

After repair work, horizontal vibration of motor NDE bearing gradually decreased in 1 year from 3.3 to 1.3 mm/s (see Figure 18.17).

Post-maintenance vibration spectrums (depicted in Figures 18.18 and 18.19) did not show any abnormality like foundation damage or looseness.

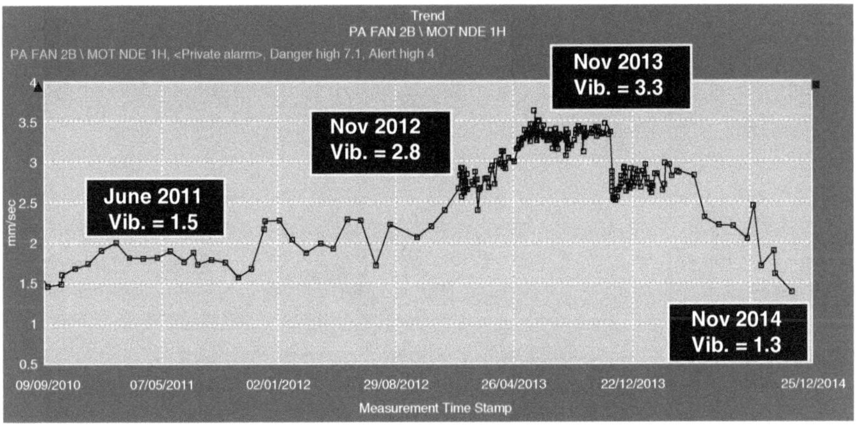

FIGURE 18.17 Post-maintenance motor NDE bearing vibration trend (horizontal).

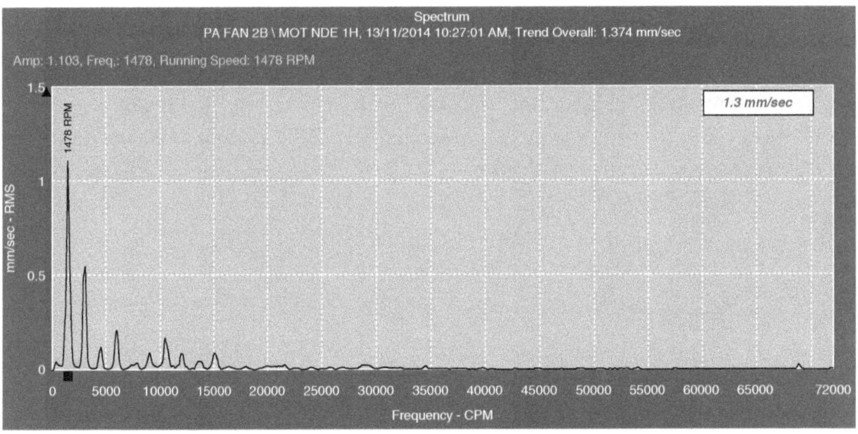

FIGURE 18.18 Post-maintenance vibration spectrum of motor NDE bearing (horizontal).

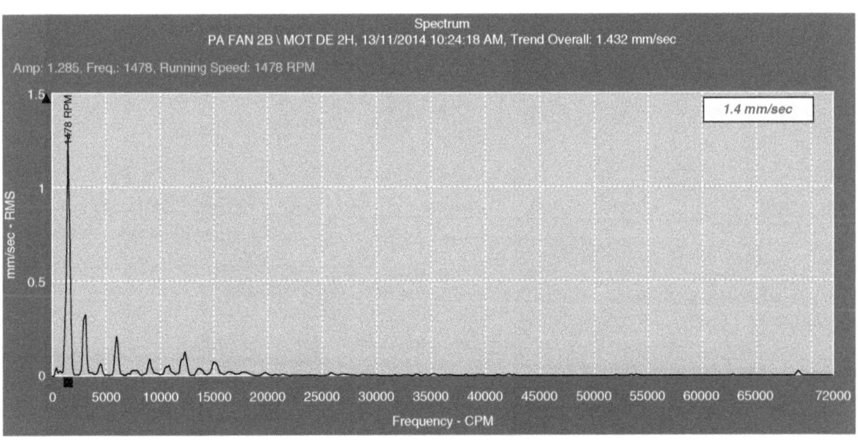

FIGURE 18.19 Post-maintenance vibration spectrum of motor DE bearing (horizontal).

18.12 CONCLUSIONS

PA fan is critical equipment which has to perform two functions (i.e. drying the coal and transportation of coal to the furnace). The boiler is provided with two nos. of primary air fans. For safe operation of each PA fan is provided with blade pitch control for controlling the loading on fans, outlet damper for isolation to facilitate start-up/ maintenance of fan.

Since PA fan has no standby, reliability of fan needs to be very high. It is a critical auxiliary in the process, if it fails, it can directly affect the generation loss of 0.09 MUs/h which costs revenue of around Rs. 315,000 per hour. If any fan motor fails during operation, it will lead to generation loss and in extreme cases tripping of plant and interruption in power supply to our valued customers in Mumbai, India.

The present case study has clearly demonstrated that condition-based vibration testing is a vital component in rotating machinery analysis and detection of abnormality at an early stage. Ultrasonic pulse velocity test, re-bound hammer test and chemical test of concrete played a major supportive role in confirming the deterioration of motor foundation well in advance so that repair work was planned systematically and completed without stopping the fan.

Catastrophic failure of motor was avoided using these techniques and also other major benefit includes prevention of downtime, avoidance of major motor repair or replacement costs. The importance of vibration and NDT techniques helped in diagnosing the exact root cause of abnormity well in advance which could have caused total cost of Rs 2,00,000.

19 Case Study 18 *Condensate Extraction Pump 2B: Pump Resonance – August 2014*

EXECUTIVE SUMMARY

The present case study is a success story of the implementation of condition monitoring techniques at ADTPS, wherein an imminent resonance problem of HT auxiliary Condensate Extraction Pump – 2B was diagnosed. The present study is a crucial one that informs us of the benefits of condition-based vibration testing. It is a vital component in rotating machinery analysis and detection of faults long before they become critical, enabling repair to be scheduled for planned production shutdowns. Bump test analysis played a major supportive role in pinpointing the resonance problem of the pump. This has saved the plant from huge losses in terms of generation, unplanned breakdown time and associated maintenance cost, which could have caused a total cost of approximately Rs. 291,000.

On 17 July 2014, after the replacement of the mechanical seal, the pump was again put back in service at full speed of 1500 rpm. On 22 July 2014, during routine monitoring, it was observed that vertical vibrations of the motor DE bearing deviated from 0.6 to 3.1 mm/s between the speed range of 1310 and 1335 rpm. From vibration spectrum analysis, it was suspected electrical/mechanical resonance problem. Variable Frequency Drive (VFD) – Threshold (THD) measurements were carried out to check any electrical resonance due to harmonics present, but all readings were found within limits which eliminated the possibility of electrical resonance. For identifying the mechanical resonance, a bump test was carried out which showed the natural frequency of the system is found 1320 rpm which is nearly close to the operating speed of the pump. After analyzing all the above databases, it was noticed that the vibration of the motor increased after mechanical seal replacement work. Hence, it was concluded that the rise of vibration might be happening due to disturbance in the contact area of the new seal with respect to the old seal which changed the stiffness of the pump and ultimately shift of natural frequency near to operating speed. This creates resonance in the pump and resulting high vibration to the motor. Due to the increased trend in vibration readings, it was decided to change over the pump. This decision was taken to re-assembly of mechanical seal of the pump and rectification was carried out accordingly.

19.1 PRE-MAINTENANCE HISTORY

Rating for motor of Condensate Extraction Pump (CEP) is 650 KW, 1500 rpm and 71.2 Amp. In 2007, variable frequency drive (VFD) was installed and generally runs at 1320 rpm for energy conservation purpose. Since 2007, CEP has been running normally, and all vibrations were stable and found in the range between 0.7 and

DOI: 10.1201/9781003432616-19

0.9 mm/s at the operating speed of 1320 rpm. Vibration trend of motor DE bearing (vertical) at a speed of 1320 rpm from 4 September 2013 to 9 July 2014 was found in a similar range (refer to Figure 19.1).

On 16 July 2014, pump outage was taken for the replacement of mechanical seal because it was having a water leakage. On 17 July 2014, after the replacement of mechanical seal, pump again put back in service at full speed of 1500 rpm due to the non-availability of VFD drive for cell replacement. At 1500 rpm, motor drive end (DE) bearing vertical vibration readings were stable at 0.6 mm/s. On 22 July 2014, VFD drive was taken in service and found that only motor DE bearing vertical vibrations were deviated from 0.6 to 3.1 mm/s at a speed of 1310 rpm (Figure 19.2).

Offline motor vibration readings were taken at various different speeds.

From the above data (shown in Figure 19.3), it was observed that the highest vibration was found in between the speed range of 1310 and 1335 rpm and all readings were found within the limit for speed more than 1340 rpm. Problem of resonance was suspected. To confirm the exact root cause of abnormality, motor was decoupled, and trial was taken.

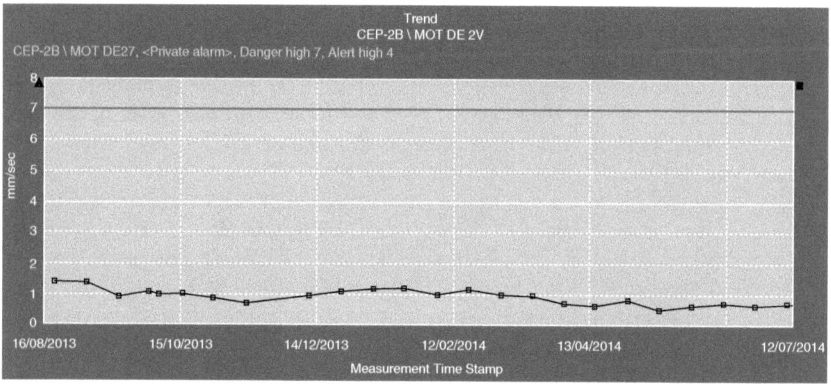

FIGURE 19.1 Vibration trend of motor DE vertical at speed of 1320 rpm.

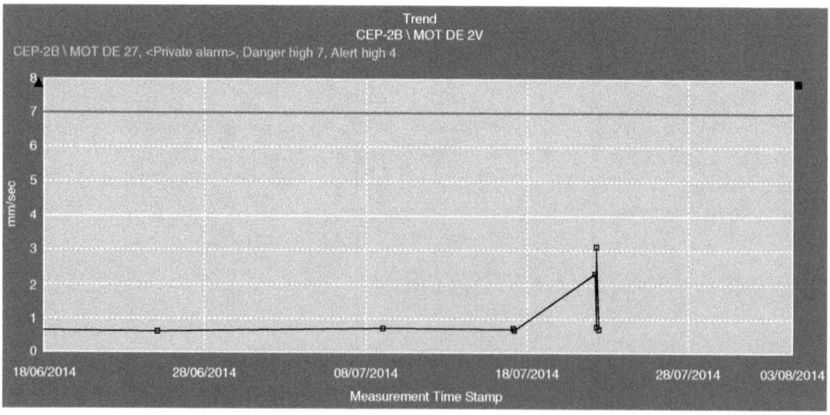

FIGURE 19.2 Vibration trend of motor DE vertical at speed of 1310 rpm.

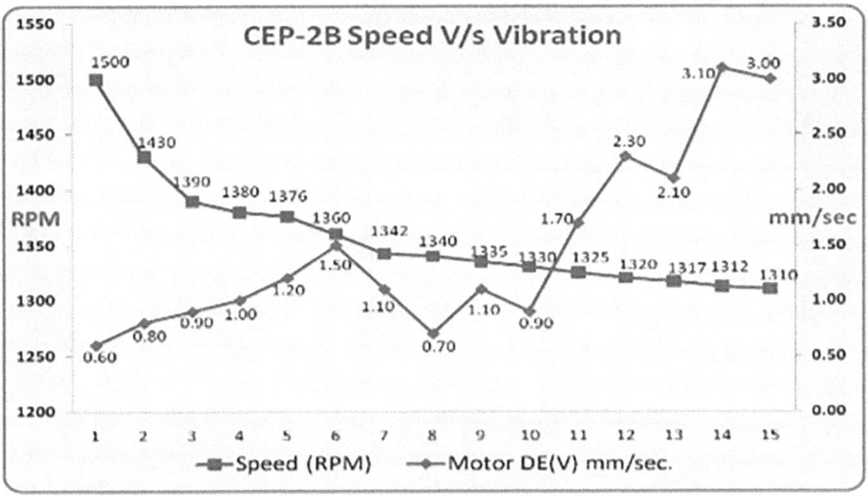

FIGURE 19.3 Motor DE bearing vibration (vertical) at various speed (in coupled condition).

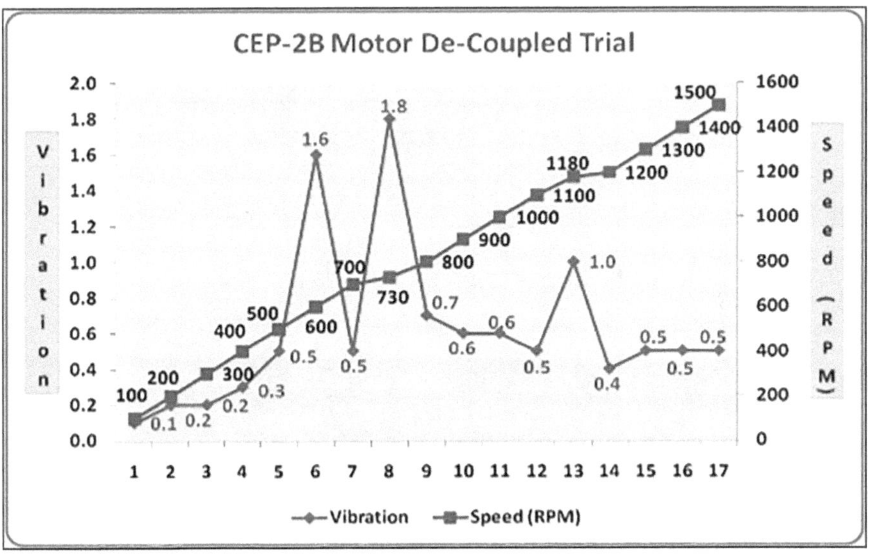

FIGURE 19.4 Motor DE bearing vibration (vertical) at various speed (de-coupled condition).

From the above data (shown in Figure 19.4), it was observed that the highest vibration was found in between the speed range of 600–750 rpm, and all readings were found within the limit at the operating speed of 1320 rpm. VFD threshold (THD) measurement of CEP motor was carried out to check any electrical resonance due to harmonics present in the system which can be responsible for this phenomenon, but all recorded readings (Figure 19.5) were found well within limits. This eliminated the possibility of the presence of electrical resonance.

Speed		Volt				Current		
		THD	3rd	5th	7th	3rd	5th	7th
1500	R	0.4	0.1	0.1	0.1	0.6	1.0	0.4
	Y	0.3	0.1	0.1	0.1	0.7	0.8	0.4
	B	0.3	0.1	0.1	0.1	0.9	1.0	0.4
1470	R	0.4	0.1	0.1	0.1	0.7	1.0	0.4
	Y	0.3	0.1	0.1	0.1	0.7	1.0	0.4
	B	0.3	0.1	0.1	0.1	0.9	1.0	0.4
1440	R	0.3	0.1	0.1	0.1	0.9	1.1	0.4
	Y	0.3	0.1	0.1	0.1	1.0	1.0	0.4
	B	0.3	0.1	0.1	0.1	1.1	1.0	0.1
1410	R	0.3	0.1	0.1	0.1	0.9	1.1	0.4
	Y	0.3	0.1	0.1	0.1	0.9	1.0	0.5
	B	0.3	0.1	0.1	0.1	1.1	1.0	0.5
1380	R	0.4	0.1	0.1	0.1	0.9	1.1	0.5
	Y	0.3	0.1	0.1	0.1	0.9	1.0	0.5
	B	0.3	0.1	0.1	0.1	1.1	1.0	0.5
1350	R	0.3	0.1	0.1	0.1	0.9	1.1	0.5
	Y	0.3	0.1	0.1	0.1	0.9	1.1	0.5
	B	0.3	0.1	0.1	0.1	1.1	1.0	0.5
1320	R	0.3	0.1	0.1	0.1	1.1	1.1	0.5
	Y	0.3	0.1	0.1	0.1	1.1	1.1	0.5
	B	0.3	0.1	0.0	0.1	1.3	1.1	0.5
1300	R	0.4	0.1	0.1	0.1	1.2	1.1	0.5
	Y	0.3	0.1	0.1	0.1	1.2	1.1	0.6
	B	0.3	0.1	0.1	0.1	1.3	1.1	0.5

FIGURE 19.5 THD measurement of CEP- 2B VFD on 23 July 2014.

FIGURE 19.6 Online motor DE bearing vibration (vertical).

In addition, pump-related parameters (i.e., pressure, temperature) were found normal. It was decided to keep a close watch on motor vibration readings by installing temporary online vibration system (Figure 19.6).

By observing such increase in trend of motor vibration reading, it was decided to change over the pump.

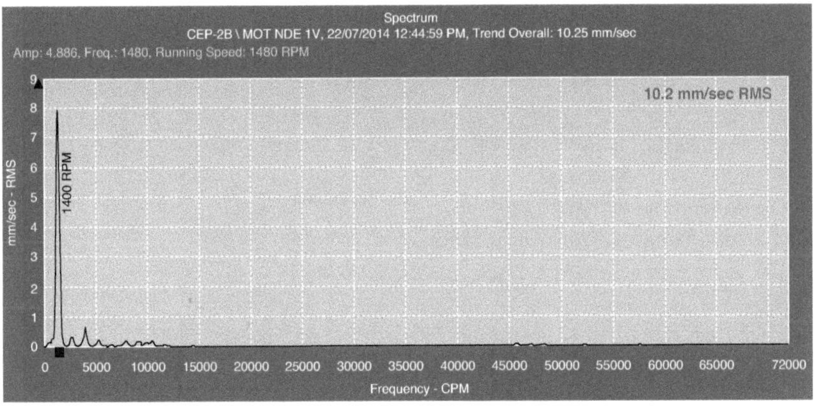

FIGURE 19.7 Spectrum of motor NDE bearing (vertical).

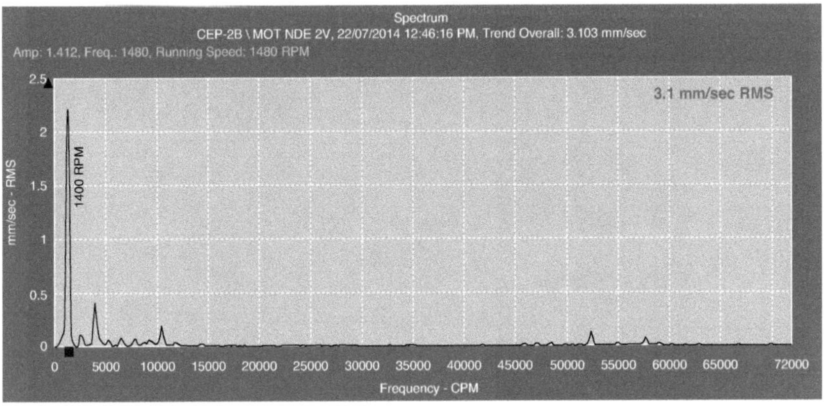

FIGURE 19.8 Spectrum of motor DE bearing (vertical).

19.2 VIBRATION DIAGNOSIS

On 22 July 2014, vertical vibrations of motor NDE and DE bearing were recorded at 10.2 and 3.1 mm/s, respectively having a speed range of 1312–1335 rpm. The same readings remained normally in the range of 2.1 and 0.6 mm/s. The spectrum analysis for root cause was investigated as follows:

From Figures 19.7 to 19.8, the predominant frequency of vibrations was found 1× rpm. Hence, the existence of mechanical resonance problem was suspected. To confirm mechanical resonance problem, bump test was carried out in the running condition.

19.3 BUMP TEST

Bump test is a simple method for analyzing the structural modal response of a machine or a structure. When any object bump/impact under this test, it needs to excite all of the natural frequencies of that object. When these frequency components coincide with the structural natural frequencies, then resonant conditions are present which result in a higher-than-normal vibration level at those frequencies (Figure 19.9).

Bump test is carried out in the following conditions:

- To excite and measure natural frequency of the object,
- To identify a resonance,
- To understand a change in mass,
- To understand a change in stiffness,
- To understand a change in damping.

Bump test technique involves simply hitting the structure with a suitable impacting device, such as a wooden mallet, a hard rubber mallet, or in some cases, a sledgehammer. This test can be performed in equipment on condition when rotor rotates where the force excitation source exists in the whole vibration system. A special algorithm called negative averaging is applied in this test (refer to Figure 19.10).

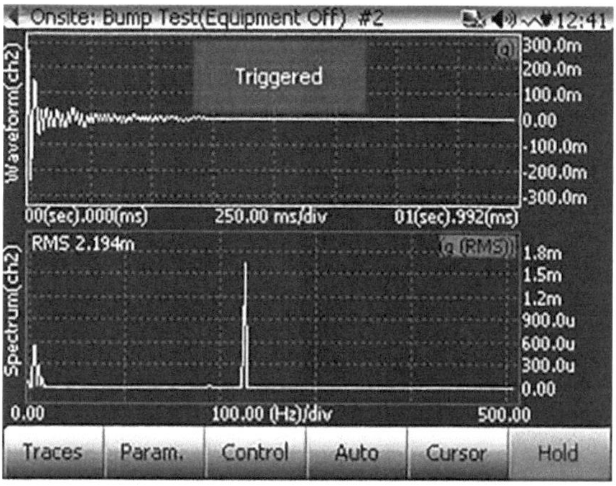

FIGURE 19.9 Fundamentals of bump test.

FIGURE 19.10 Bump test carried out for CEP motor in running condition.

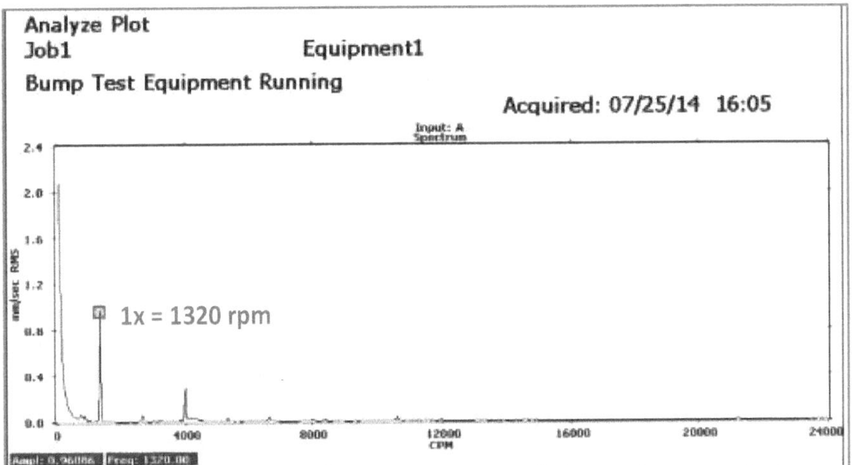

Frequency	Data	Frequency	Data	Frequency	Data	Frequency	Data
465.2	0.07499	731.2	0.07335	844.0	0.06627	976.3	0.01753
1066.6	0.01372	1326.3	0.98403	2652.0	0.06513	3004.6	0.01473
3223.8	0.01561	3978.3	0.32381	4185.1	0.03642	4306.0	0.03086
4514.6	0.01819	5346.0	0.03652	6636.4	0.04475	7992.9	0.01731
8355.2	0.0155	10608.4	0.04256	11934.9	0.01359	21221.4	0.01493

FIGURE 19.11 Bump test spectrum.

Analysis shown in Figure 19.11 revealed that the natural frequency of the system was found 1320 rpm which was very close to the operating speed of pump (i.e. 1310 rpm). It clearly pinpoints the existence of mechanical resonance in pump.

19.4 BUMP TEST ANALYSIS

It was observed that there was an increase in the amplitude of vibrations for motor NDE and DE bearing only in the vertical direction as recorded at 10.2 and 3.1 mm/s, respectively having a speed range of 1312–1335 rpm. Earlier, it was found in the range of 2.1 and 0.6 mm/s. respectively. Spectrum analysis revealed that there was an increase in 1× frequency. Thus, it was suspected the problem of mechanical/electrical resonance. THD measurement of VFD was carried out to check any electrical resonance due to harmonics present in the system which can cause this resonance. All readings were found normal which eliminated the possibility of electrical resonance due to VFD harmonics. To confirm mechanical resonance, bump test was carried out. Bump test analysis reflected that natural frequency of the system was 1320 rpm which was very close to operating speed of the pump. This condition clearly pinpointed the existence of mechanical resonance in the pump.

With reference to mentioned all above analysis, it was noticed that vibration of motor increased after mechanical seal replacement work which was done on 16 July 2014. Hence, it was concluded that rise of vibration might be happening due to disturbance in contact area of new mechanical seal with respect to old mechanical seal which might has change the stiffness of pump and ultimately shifted of natural frequency near to operating speed. This created a mechanical resonance in pump. Decision was taken to re-assemble pump mechanical seal.

19.5 PERFORMED MAINTENANCE WORK

Motor de-coupled from the rest of the group drives (see Figure 19.12). Before dismantling, initial alignment readings were taken for reference. Pump bearing housing was removed and its components were dismantled as shown in Figure 19.13. Mechanical seal was removed. Inspection and cleaning were performed and completed (see Figure 19.14). Re-fitting of mechanical seal was completed. Bearing assembly box up, fitting along with pump float adjustment done (see Figure 19.15). Motor installation and alignment were completed. CEP trial was taken at full load and vibration readings were found well within the limit between the speed range of 1312 and 1335 rpm (Figure 19.16).

Bump test was carried out and found that the natural frequency shifted from the operating zone (i.e. 1320–840 rpm). Pump boxed up and made it available for operation.

FIGURE 19.12 Motor decoupled from pump.

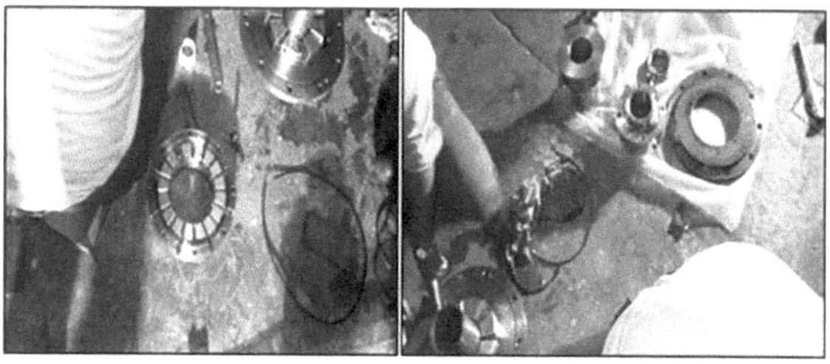

FIGURE 19.13 Removal of pump bearing.

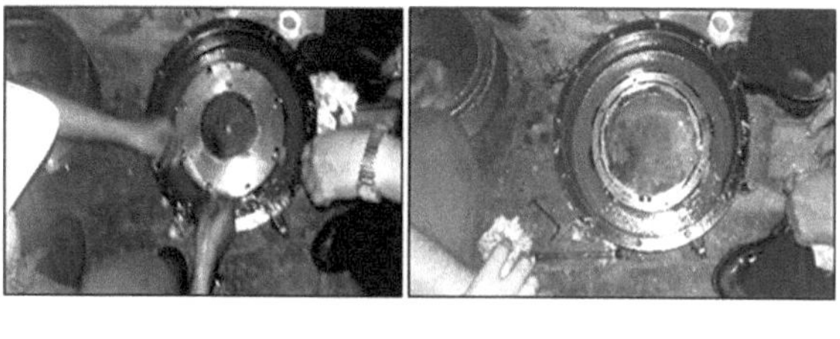

FIGURE 19.14 Inspection and cleaning of mechanical seal.

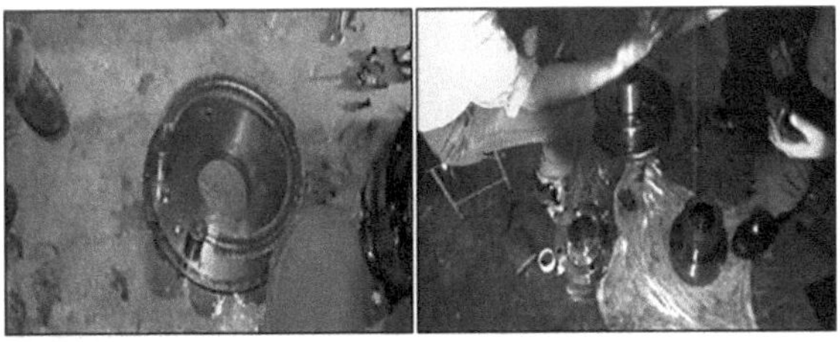

FIGURE 19.15 Re-assembly of mechanical seal.

FIGURE 19.16 Post-maintenance online vibration of motor DE bearing (vertical).

19.6 POST-MAINTENANCE OFFLINE VIBRATION READINGS

Table 19.1 represents a comparison of vibration data for pre- and post-maintenance. All values were reduced to normal level after mechanical seal re-fitting. Figures 19.17–19.19 represent the post-maintenance bump test and vibration spectrum of the motor which shows shifting of natural frequency resulting in low vibration level. This concludes the success of maintenance work execution.

TABLE 19.1
Vibration Report of CEP Pump 2B

			Initial Readings	After Re-assembly of Mechanical Seal
			22/07/2014	09/08/2014
Point		Position	Velocity (mm/s) RMS	
Motor	(NDE)	H	8.0	3.2
		V	**10.2**	**2.4**
		A	2.3	1.3
Motor	(DE)	H	0.6	0.5
		V	**3.1**	**0.7**
		A	2.3	0.8
Pump	(DE)	H	1.0	0.5
		V	0.7	0.7
		A	0.8	0.6
Natural frequency (*using Bump test*)			1320 rpm	840 rpm

19.7 POST-MAINTENANCE BUMP TEST SPECTRUM

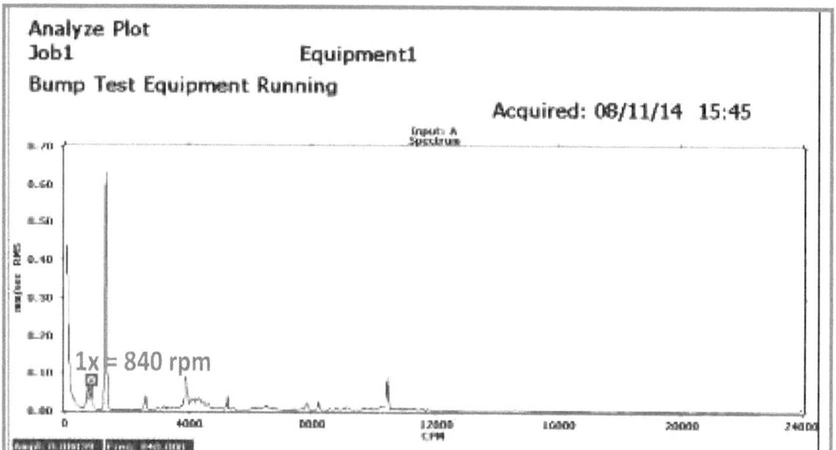

Frequency	Data	Frequency	Data	Frequency	Data	Frequency	Data
258.1	0.04296	375.4	0.02125	435.7	0.01569	735.7	0.08071
846.4	0.08218	1307.8	0.70809	2617.7	0.04292	3253.7	0.01539
3940.0	0.09606	4153.2	0.03976	4272.8	0.03742	4362.4	0.03728
4513.0	0.02732	4667.8	0.0211	5271.3	0.04194	6548.3	0.02123
7847.3	0.02415	8244.6	0.02873	10218.2	0.01636	10465.0	0.097

FIGURE 19.17 Post-maintenance bump test spectrum.

19.8 POST-MAINTENANCE VIBRATION SPECTRUMS

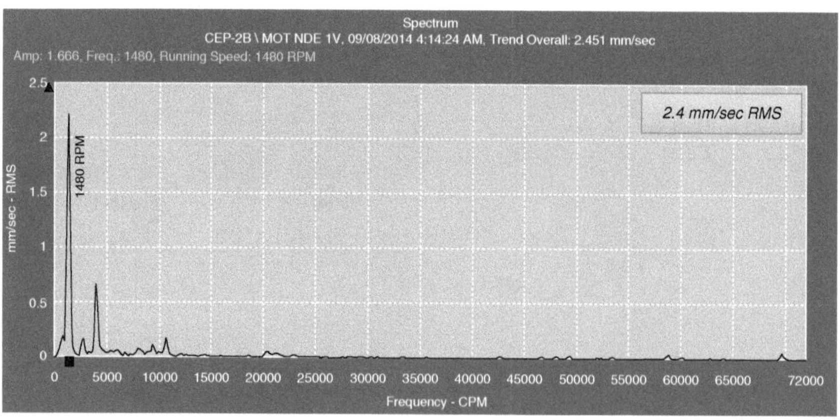

FIGURE 19.18 Post-maintenance vibration spectrum of motor NDE bearing (vertical).

Post-maintenance vibration spectrums did not show any abnormality such as misalignment or looseness.

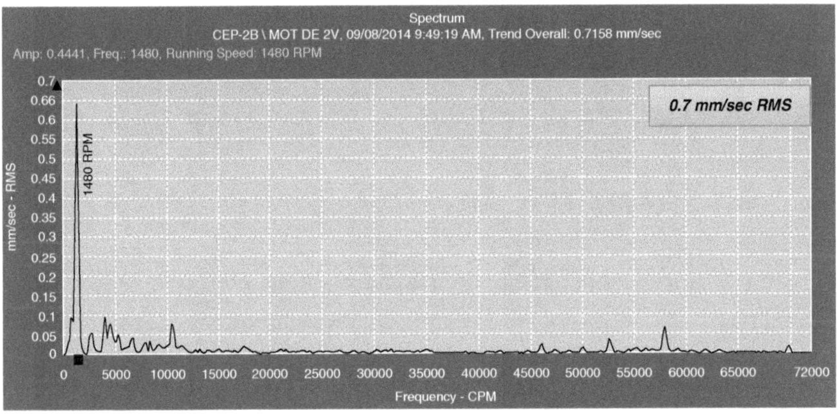

FIGURE 19.19 Post-maintenance vibration spectrum of motor DE bearing (vertical).

19.9 CONCLUSIONS

CEP is a critical auxiliary used in any thermal power plant for feeding condensed steam back to the boiler. Therefore, the reliability of the pump needs to be very high. There are two nos. of CEP of 100% capacity each (i.e. one running at full load and the other is in stand-by condition). If this pump fails and the stand-by pump is unavailable due to maintenance, it will lead to a sudden rise in the hot well condensate level which will deteriorate the condenser vacuum and ultimately trip or lead to a forced outage of the unit. This unplanned outage will hamper the reliability of the power supply to customers.

The present case study represents the benefits of condition-based vibration testing. It is a vital component in rotating machinery analysis and detection of faults long before they become critical, enabling repair to be scheduled for planned production shutdowns.

Bump test analysis played a major supportive role in pinpointing the resonance problem of pump by identifying the presence of natural frequency close to operating speed resulting in high vibration to motor. The catastrophic failure of motor and pump bearing is avoided using these techniques and also other major benefit includes prevention of downtime, avoidance of major motor and pump repair or replacement costs. Vibration and bump test measurements helped in diagnosing the exact root cause of abnormity well in advance which could have caused a total cost of Rs 2,91,000.

20 Case Study 19
ID Fan Motor Trial (at Test Bench): Motor Bearing Damage – June 2014

EXECUTIVE SUMMARY

The present case study is a success story of the implementation of condition monitoring techniques at ADTPS, wherein imminent bearing failure of Spare HT auxiliary ID Fan Motor was diagnosed at no load de-coupled condition on test bench.

Vibration analysis and distress level measurement helped in diagnosing the exact root cause of abnormity of motor bearing at early stage and thus avoided the catastrophic failure. This has saved the plant from huge losses in terms of generation, unplanned breakdown time and associated maintenance cost, which could have caused a total cost of approx. Rs. 2.00 lakhs.

At ADTPS, electrical maintenance and condition monitoring cell has adopted proactive practice of collecting the vibration and acceleration database at no load de-coupled condition on the dedicated test bench after any HT/LT motor overhaul/servicing for knowing the success of maintenance work.

On 19 March 2014, after motor overhauling, trial run of de-coupled condition was carried out and observed that the acceleration value of motor DE and NDE bearing was found in the range of 1.7 and 1.1 g which was on higher side (as normally it ranges below 1.0 g). It indicated bearing deterioration at the initial stage although both DE and NDE bearing vibration and temperatures were observed normal. To confirm, the fault bearing distress level was checked and found on higher side of level 10. Meanwhile, motor greasing was carried out and the matter was taken up to OEM.

On 14 May 2014, again trial was taken in the presence of OEM representative, but still acceleration value found was on higher side at 2.7 and 1.5 g. In view of the increasing trend of bearing acceleration and distress level, it was decided to replace the bearings of both motor DE and NDE.

Bearing failure analysis was done by OEM (M/S SKF India Ltd), and their report stated that a symmetrical scratch mark that was observed on roller is due to the metallic particle which may be entrapped between the roller and cage pocket. Such kind of mark is only possible when there is any particle entrapped between the mating surfaces. The acceleration value will also rise due to such marks. Detailed inspection of rotor and stator winding of motor was carried out and no damage was observed.

20.1 PRE-MAINTENANCE HISTORY

On 19 March 2014, after motor overhauling, trial run of ID fan motor was carried out for 20 min at no load de-coupled condition. During the trial, it was observed that acceleration value of motor drive end (DE) and non-drive end (NDE) bearing was

DOI: 10.1201/9781003432616-20

observed on higher side and reported as 1.7 (peak) and 1.1 g (peak), respectively. At the same time, no deviations were observed in temperature and vibrations of motor DE and NDE bearing. Meanwhile, it was decided to grease the motor bearings and matter should be taken up to bearing OEM manufacture. On 14 May 2014, again trial was taken in the presence of OEM representative, but the acceleration value of motor DE and NDE bearing was found on higher side and reported as 2.7 and 1.5 g, respectively (refer to Figure 20.1). In addition, bearing distress level was checked and found on higher side of level 10 dst. In view of increasing trend of bearing acceleration and distress level, it was decided to replace both motor DE and NDE bearings.

20.2 VIBRATION DIAGNOSIS

The acceleration values of motor DE and NDE bearing were recorded as 2.7 and 1.5 g, respectively at no load while, at same load reading never exceed beyond 2.0 g. Figures 20.2–20.5 discuss about vibration and acceleration spectrum of motor DE

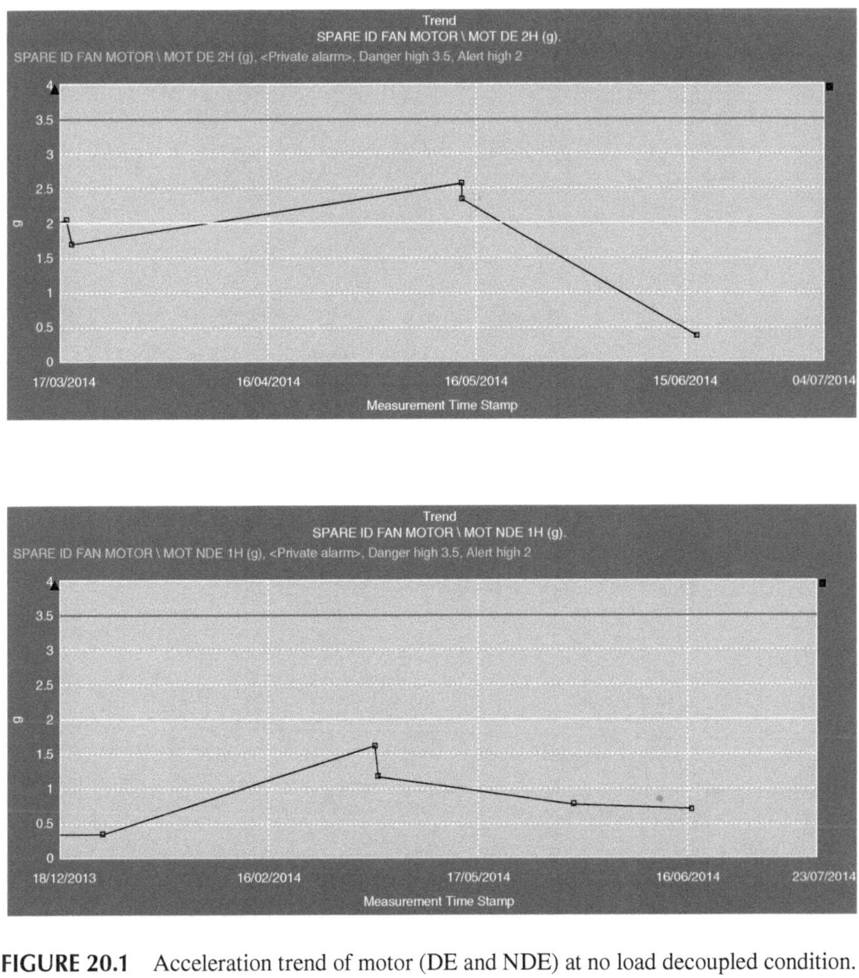

FIGURE 20.1 Acceleration trend of motor (DE and NDE) at no load decoupled condition.

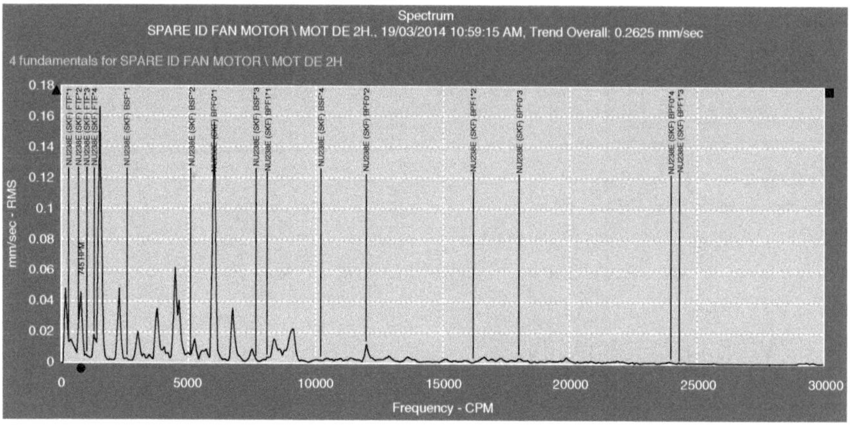

FIGURE 20.2 Vibration spectrum of motor DE bearing (horizontal).

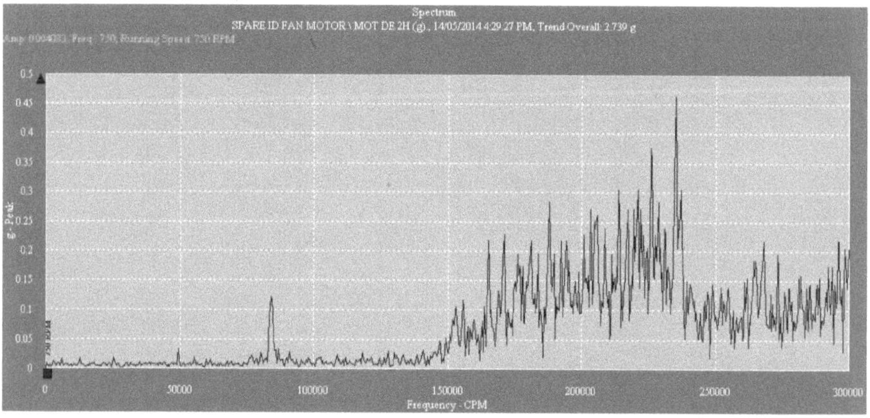

FIGURE 20.3 Acceleration spectrum of motor DE bearing.

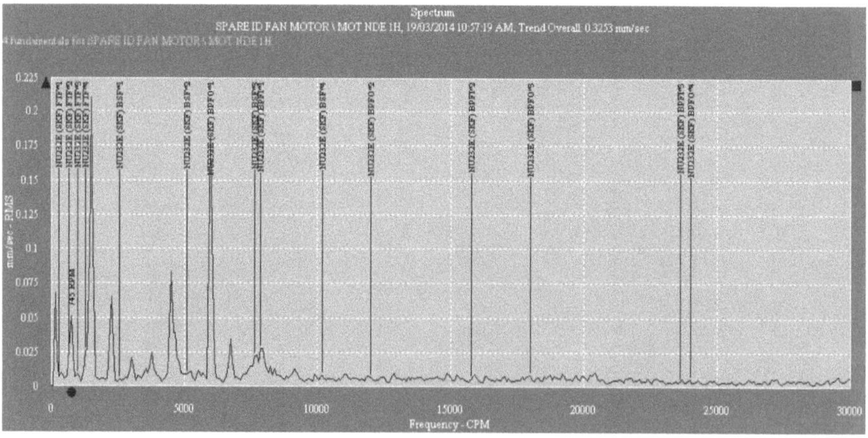

FIGURE 20.4 Vibration spectrum of motor NDE bearing (horizontal).

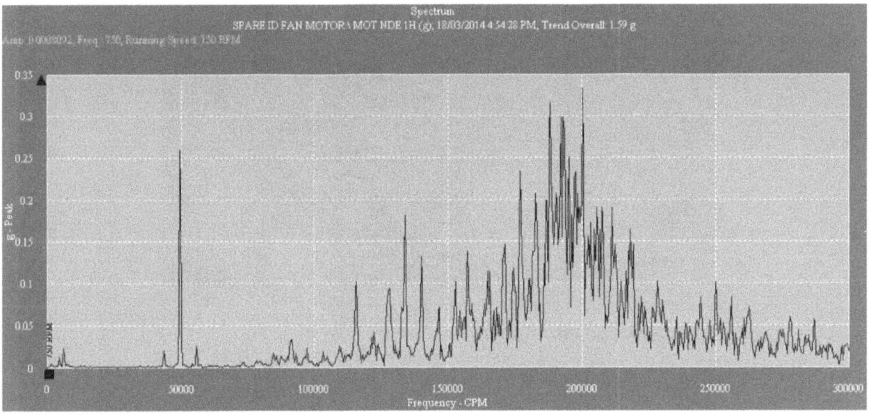

FIGURE 20.5 Acceleration spectrum of motor NDE bearing.

and NDE bearing showing defect frequencies at Ball Pass Frequency Outer race (BPFO) and Ball Pass Frequency Inner race (BPFI).

The spectrum analysis for root cause was investigated as follows:

20.3 DISTRESS LEVEL MEASUREMENT USING ADVANCE BEARING CHECKER

Working principles for this process are as follows:

- When mechanical condition of machinery (such as bearings) deteriorates, energy loss processes (such as impacts, friction and crushing) generate sound wave activity that spans a broad range of frequencies. By detecting only, the high frequency part of this signal with special acoustic emission (AE) sensors, it is possible to detect miniscule amounts of activity (e.g., a slight rub, a brief impact, or the crushing of a single particle in the lubricant).
- When unit through the magnetic sensor head is attached to bearing casing, within 10 seconds both dB level and distress values will be displayed. dB level is an indication of the overall noise of the bearing and is dependent on speed. It increases not only with increase in speed of rotation but also with degradation of the bearing or inadequate lubrication.
- Distress value gives an instant indication of the state of the bearings' health. A reading below 10 generally indicates normal operation; higher than 10 is usually indicative of bearing damage or the need for attention (refer to Figure 20.6).

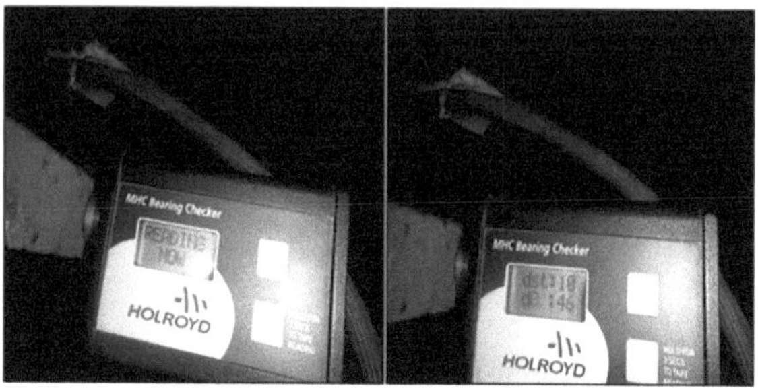

FIGURE 20.6 Distress level data collection.

20.4 FREQUENCY ANALYSIS

Initially, acceleration values of motor DE and NDE were observed on higher side (2.7 and 1.5 g, respectively). Normally, it does not exceed 1.0 g at no load de-coupled condition. Therefore, it was suspected that motor bearing deterioration was in progress. In addition, by noticing the bearing distress level as 10 dst which indicates the bearing damage. Vibration spectrums show bearing defect frequencies at BPFO and BPFI which clearly pinpoints bearing defects. Acceleration spectrums revealed that the contribution of non-synchronous energy was at higher side which was mainly due to bearing defects. These frequency stress waves' events occur at repeatable time intervals corresponding to a bearing defect frequency, then it proved that a bearing defect was present. Hence, it was concluded that the acceleration and distress level of both motor bearings increased due to deterioration of bearings and early damage was in progress. So, it was decided to stop the motor for bearing replacement.

20.5 OBSERVATIONS DURING MAINTENANCE WORK

Condition of cooling fan and both bearings' housings were found okay. Detailed bearing analysis was carried out by OEM M/s SKF India Limited. OEM report stated that symmetrical scratch mark observed on the roller is due to the metallic particle which may be entrapped between the roller and cage pocket. Such mark is only possible when there is any particle entrapped between the mating surfaces. The acceleration value can also rise due to such marks. A scratch mark observed on the rollers center which was symmetrical over the complete circumference of the roller (shown in Figure 20.7).

FIGURE 20.7 Scoring marks on bearing roller.

20.6 PERFORMED MAINTENANCE WORK

Initially, cooling fan from NDE side was removed. Both DE and NDE bearings were removed and replaced with the new one. Then, cooling fan was mounted on motor shaft. Motor trial was taken at no load in decoupled condition. Vibration and acceleration values were found in well within limit.

20.7 POST-MAINTENANCE OFFLINE VIBRATION READINGS

Table 20.1 and Figures 20.8 and 20.9 represent the comparison of vibration, acceleration and distress level data for pre- and post-maintenance along with their spectrums. All values reduced to normal level after replacing motor DE and NDE bearings. This concludes the success of maintenance work executed by electrical team.

TABLE 20.1
Vibration Report of Spare ID Fan Motor

			Initial Readings		After Replacing Bearings	
			14/05/2014		17/06/2014	
Point		Position	Velocity (mm/s) RMS	Acceleration Value (g)	Velocity (mm/s) RMS	Acceleration Value (g)
Motor	(NDE)	H	0.4	**1.5**	0.3	**0.6**
		V	0.2	-	0.2	-
		A	0.3	-	0.2	-
Motor	(DE)	H	0.4	**2.7**	0.3	**0.3**
		V	0.3	-	0.3	-
		A	0.3	-	0.2	-
Motor	(DE)	H	Distress level = **10 dst**		Distress level = **05 dst**	

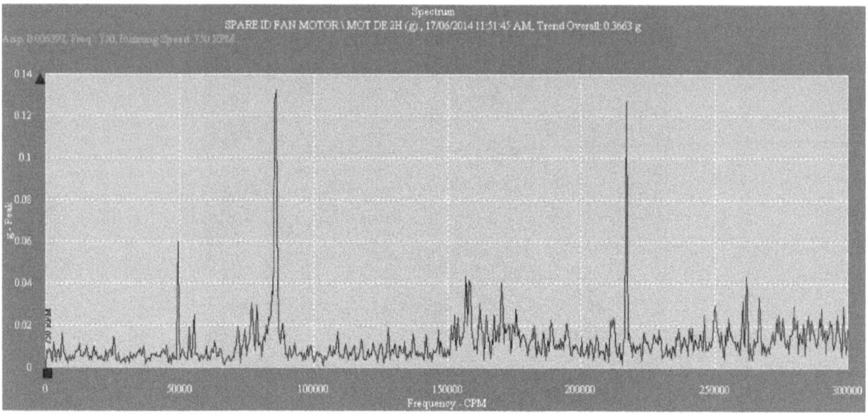

FIGURE 20.8 Acceleration spectrum of motor DE.

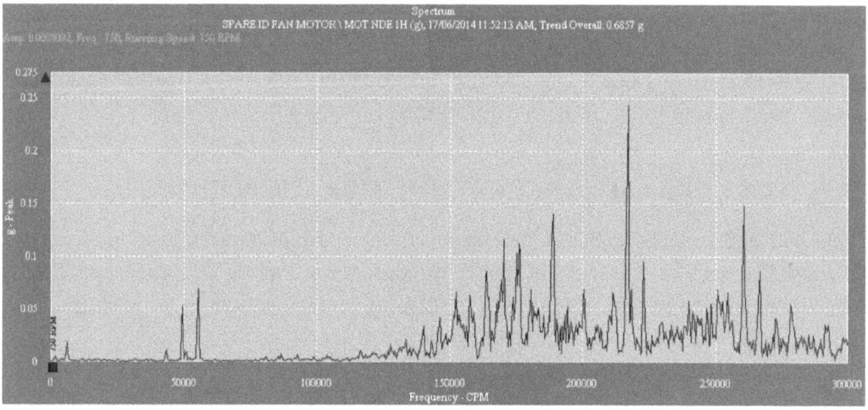

FIGURE 20.9 Acceleration spectrum of motor NDE.

20.8 POST-MAINTENANCE VIBRATION SPECTRUMS

From post-maintenance spectrums, no abnormality such as bearing defect frequencies was observed after replacement of motor bearings.

20.9 CONCLUSIONS

ID fan is critical equipment which is used to extract the flue gases out of furnaces via economizer, air preheater through stack to atmosphere. It assists the FD fan by drawing out combustible gases from the furnace and maintains the partial negative pressure inside the Boiler. Since ID fan has no standby, the reliability of fan needs to be very high. Boiler is provided with two numbers of induced draft fans. Each ID fan is provided with regulating damper control and scoop control for controlling the loading on fans. This made ID fan critical equipment in the system. If any fan motor fails during operation, it will lead to generation loss and in extreme case tripping of plant and interruption in power supply to valued customers in Mumbai, India.

The present case study has clearly demonstrated that acceleration is a valuable parameter in rotating machinery analysis and is especially useful in detecting early and advanced bearing degradation even the vibration levels are on lower side. Distress level measurement also played a major supportive role to pinpoint that bearing deterioration is in progress. OEM also agreed with this analysis and provided free replacement of both bearings. The catastrophic failure of motor at no load de-coupled condition was avoided using these techniques, and other major benefits include prevention of downtime, avoidance of major motor repair or replacement costs. The importance of acceleration and distress level measurement helped in diagnosing the exact root cause of abnormity at early stage, which could have caused a total cost of Rs 2.00 lakhs.

21 Case Study 20 *HP Pump C (AHP): Motor Rotor Bar Damage – June 2014*

EXECUTIVE SUMMARY

The present case study is a success story of the implementation of condition monitoring techniques at ADTPS, wherein imminent rotor degradation due to crack in rotor bar of HT auxiliary HP Pump – C motor was diagnosed.

Motor current signature analysis (MCSA) clearly demonstrated that it is a powerful online monitoring tool for diagnosing the exact root cause of abnormity of HT auxiliary HP Pump – C motor at very early stage and thus avoided the catastrophic failure. This practice has saved the plant from huge losses in terms of generation, unplanned breakdown time and associated maintenance cost, which could have caused total cost of approximately Rs. 5.00 Lacs.

On 21 March 2014, during routine monitoring, it was observed that motor current index (MCI) deviated from 44.04 to 41.66 dB. Motor and pump vibrations and all operating parameters were found within limit. On 3 April 2014, MCI was deviated from 41.66 to 40.67 dB. It was decided to keep a close watch on MCI.

On 18 April 2014, MCI further deviated from 40.67 to 38.63 dB. In the view of decreasing trend of MCI, it was decided to stop the pump for repairing the rotor of motor. Meanwhile during the inspection of motor rotor, rotor bars were found cracked and damaged. Rotor was repaired in-house by silver brazing, and the motor was made available with this repaired rotor. Motor was taken in service at load, and MCI was found to be improved from 38.63 to 49.89 dB. During maintenance, detailed inspection of motor bearings and stator winding was performed, and no damage/scoring marks were observed.

21.1 PRE-MAINTENANCE HISTORY

On 21 March 2014, during routine monitoring, it was observed that motor current index (MCI) was deviated from 44.04 to 41.66 dB (Figure 21.1). Vibration readings of motor and pump were found within the limit (Refer Figures 21.2 and 21.3). All the operating parameters, (i.e., pressure, temperature) were found to be normal. On 3 March 2014, motor current index deviated from 41.66 to 40.67 dB. On 3 March 2014, detailed vibration, and motor current signature analysis (MCSA) report along with various probable causes were discussed in HoD meeting. It was decided to keep a close watch on motor current index. On 18 April 2014, it was observed that MCI was deviated from 40.67 to 38.63 dB. In view of decreasing trend of MCI, it was decided to stop the pump for repairing of motor rotor.

DOI: 10.1201/9781003432616-21

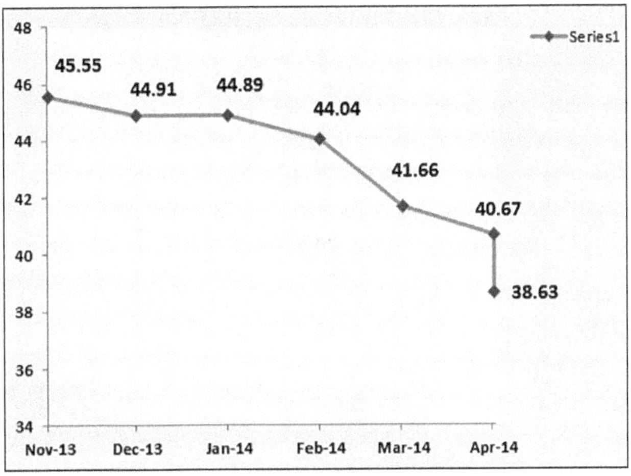

FIGURE 21.1 Monthly trend of motor current index.

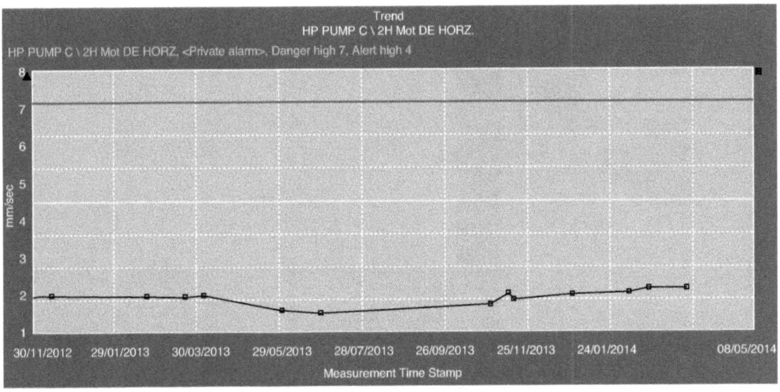

FIGURE 21.2 Vibration trend of motor DE bearing (horizontal).

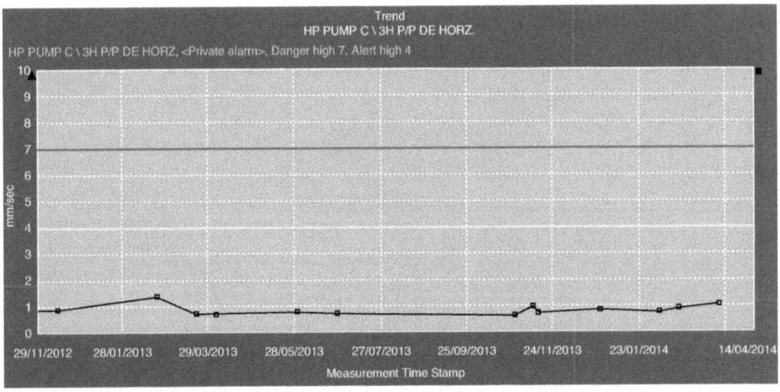

FIGURE 21.3 Vibration trend of pump DE bearing (horizontal).

21.2 FUNDAMENTALS OF MOTOR CURRENT SIGNATURE ANALYSIS (MCSA)

When the rotor bars start degrading, the rotor impedance rises, and high resistance occurs in motor rotor. Also, harmonic fluxes are produced in the air gap. This flux induces a current component in the stator winding. This current component causes modulation/fluctuation of the supply current at positive the pole pass frequency (PPF = Motor slip × No. of poles) of the motor (see Figure 21.4).

A Fast Fourier transform (FFT) tool is used for converting signals from time domain to frequency domain. In a spectrum of motor current, this modulation creates pole pass sidebands around line frequency peak (Figure 21.5).

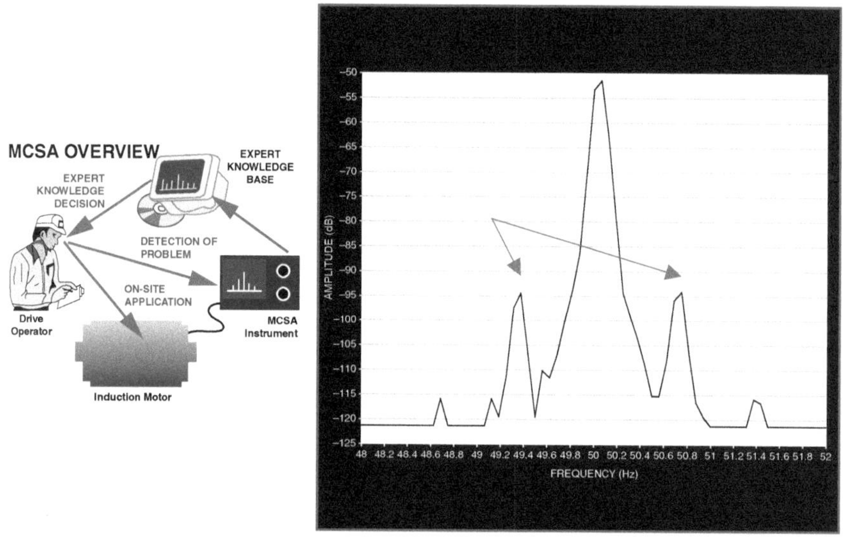

FIGURE 21.4 Fundamentals of MCSA.

Severity level	F_L / FP (dB)	F_L / F_P (Ratio)	F_P / F_L (Ratio %)	Rotor Condition Assessment	Recommended corrective action
1	>60	>1000	<0.10	Excellent	None
2	54–60	50–1000	0.10–0.20	Good	None
3	48–54	251–501	0.20–0.40	Moderate	Trend data
4	42–48	126–251	0.40–0.79	Rotor bar crack may be developing or problems with high resistance joints	Increase trending frequcency
5	36–42	63–126	0.79–1.58	One or two rotor bars likely cracked or broken	Perform vibration test to confirm source & severity
6	30–36	32–63	1.58–3.16	Multiple cracked or broken rotor bars	Repair ASAP
7	<30	<32	>3.16	Multiple cracked or broken rotor bars & end-rings	Repair or replace ASAP

FIGURE 21.5 Rotor bar severity level chart.

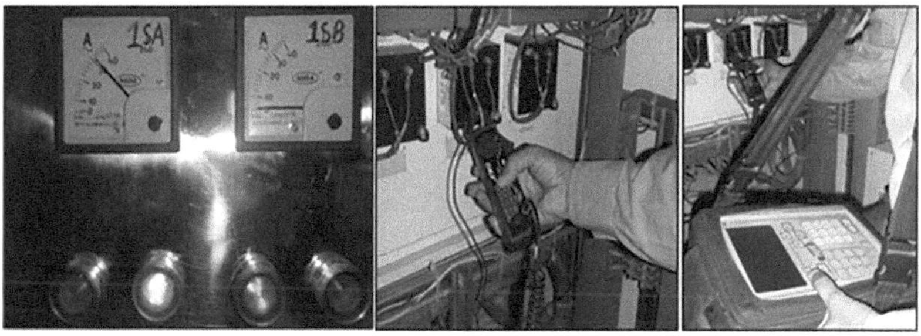

FIGURE 21.6 MCSA data collection from Motor Control Centre (MCC).

21.3 MCSA DATA COLLECTION

MCSA is capturing current and voltage signal and analyzing them to detect various electrical faults in motor. It uses the electric motor as a transducer, allowing the user to evaluate the electrical and mechanical condition from motor control centre (MCC) enabling easy testing of motors located at remote, inaccessible, or hazardous areas (refer to Figure 21.6).

21.4 MOTOR CURRENT SIGNATURE ANALYSIS

From Figures 21.7 to 21.9, on 18 April 2014, the MCI deviated from 40.67 to 38.63 dB, the same reading should remain above 45 dB at full load. The spectrum analysis for root cause was investigated as follows:

MCI calculation:

- Line frequency (LF) component = 15.90
- Sideband $A = 0.186930$
- Sideband $B = 0.141150$
- PPF component = $A = 0.186930$
- MCI = 20 log (LF/PPF)
- MCI = 20 log (15.90/0.186930)
- MCI = 38.63 dB

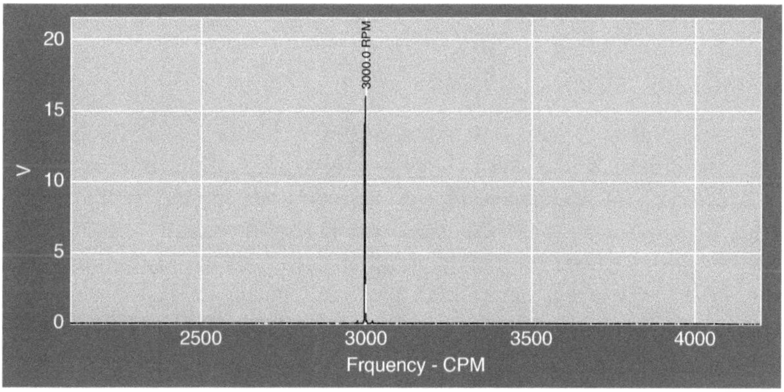

FIGURE 21.7 Motor current spectrum.

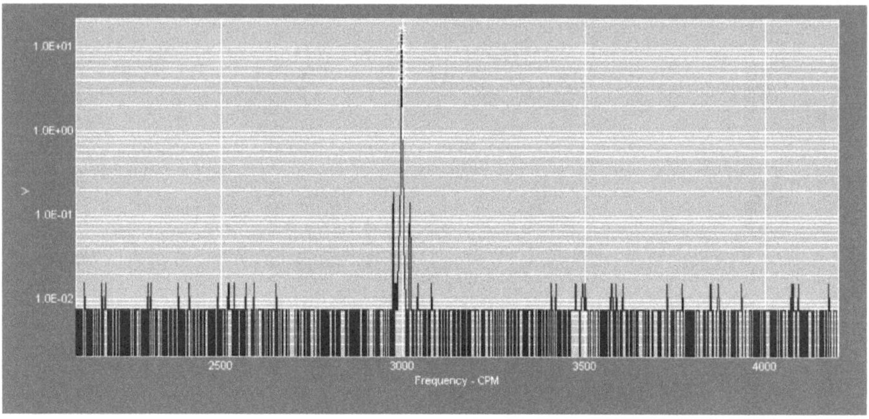

FIGURE 21.8 Motor current spectrum in log form.

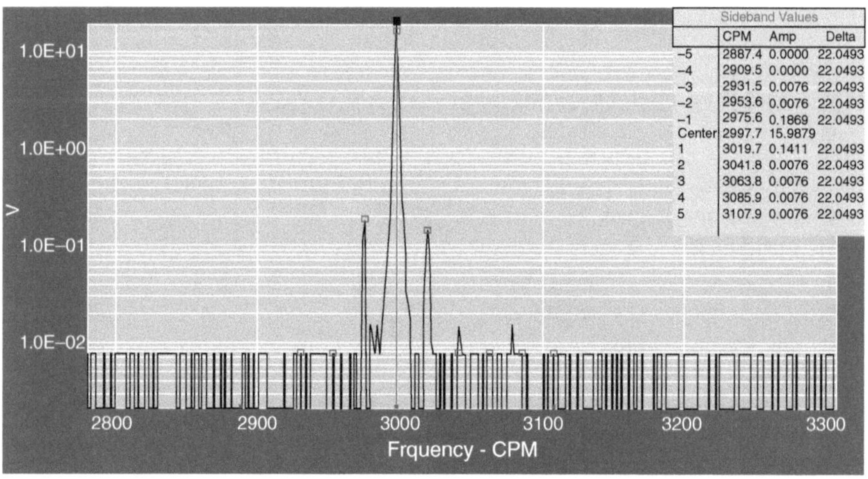

FIGURE 21.9 Zoomed motor current spectrum showing sidebands to line frequency 50 Hz.

21.5 MCI ANALYSIS

Initially motor current index was found on lower side i.e., 40.67 dB. Normally, it ranges above 45 dB at full load. Motor current spectrum analysis (refer to Figure 21.9) revealed that there was an increase in pole pass sidebands around line frequency peak (50 Hz). It was suspected that there was development of crack in rotor bar. Further, this MCI reached to lower level at 38.63 dB. Hence, it was concluded that low MCI value was due to loose or crack in rotor bar which means the deterioration of rotor was in progress. Therefore, it was decided to stop the pump for repairing motor rotor.

21.6 OBSERVATIONS DURING MAINTENANCE WORK

Condition of cooling fan, bearings and bearing housings were checked and found okay. No scoring marks were observed on both DE and NDE shaft ends. Rotor bars were found cracked and damaged (see Figure 21.10).

Detailed inspection of motor winding was carried out, and no scoring marks were observed on it as shown in Figure 21.11.

FIGURE 21.10 Damaged rotor bars.

FIGURE 21.11 Inspection of motor stator.

21.7 PERFORMED MAINTENANCE WORK

Initially, coupling and both bearings were removed from motor. Rotor of motor was repaired in-house by silver brazing (refer to Figure 21.12). Then, both DE and NDE bearings and coupling were mounted on motor shaft. Motor decoupled trial was taken, and vibration readings were found in well within limit.

Motor alignment was done with the rest of drives, boxed up and made it available for operation. Motor was taken in service at load condition, and vibration readings were found in well within limit. MCI was improved from 38.63 to 49.89 dB.

21.8 POST-MAINTENANCE RESULTS

Table 21.1 and Figure 21.13 represent the comparison of vibration, acceleration, and MCI index data for pre- and post-maintenance. MCI index improved to normal level

FIGURE 21.12 Repaired motor rotor.

TABLE 21.1
Vibration and MCI Report of HP Pump C

			Initial Readings		After Repairing Motor Rotor	
			18/04/2014		03/06/2014	
Point		Position	Velocity (mm/s) RMS	Acceleration Value (g)	Velocity (mm/s) RMS	Acceleration Value (g)
Motor	(DE)	H	1.0	0.28	0.7	0.38
		V	0.6	-	0.5	-
		A	0.4	-	0.3	-
Pump	(DE)	H	0.6	0.23	0.3	0.27
		V	0.6	-	0.4	-
		A	0.3	-	0.4	-
Motor Current Index (MCI)			**38.63 dB**		**49.89 dB**	

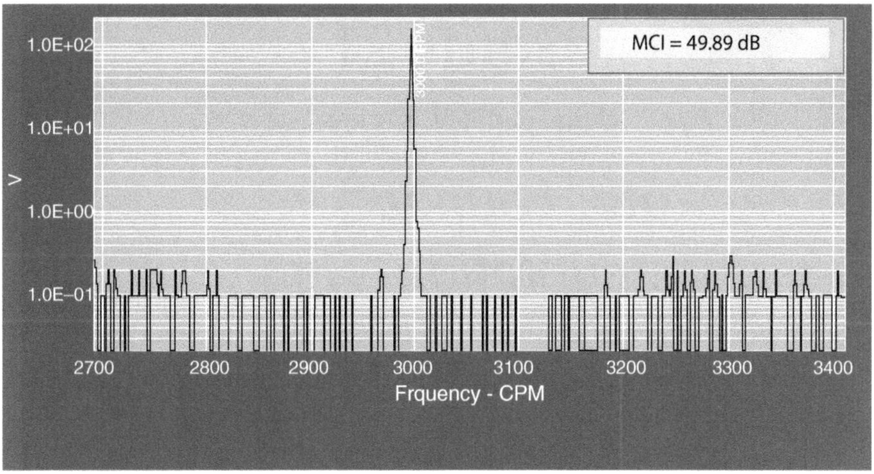

FIGURE 21.13 Post-maintenance motor current spectrum recorded on 3 June 2014.

after repairing motor rotor. This concludes the success of maintenance work executed by electrical team.

Post-maintenance motor current spectrum did not show any abnormality such as sideband frequencies around line frequency after repairing of rotor.

21.9 CONCLUSIONS

High pressure (HP) pump of ash handling plant (AHP) is a specific type of pump which is used to pump the condenser discharge water at pressure 10.5 kg/cm^2 and flow 520 m^3/h for deashing of bottom ash hopper, economizer, air pre-heater (APH) duct and electrostatic precipitator (ESP). Failure of this pump will affect the redundancy of the HP pump system that has been used for deashing process. This may lead to total stoppage of deashing process and ultimately forced outage of unit due to accumulation of ash.

This case study has clearly demonstrated that MCSA is a powerful online monitoring technique for assessing the operational condition of HP pump induction motor. The avoidance of catastrophic failure has been achieved through MCSA technique and other major benefits includes prevention of downtime, avoidance of major motor repair or replacement costs. The importance of MCSA measurement is to ensure reliable data, operational conditions of motor and correct interpretation of data which helped in diagnosing the exact root cause of abnormity at early stage that might have caused total cost of Rs. 5.00 lakhs.

22 Case Study 21 *DC Seal Oil Pump 1A: Motor Foundation Deterioration – July 2013*

EXECUTIVE SUMMARY

Following is the success story of implementation of condition monitoring techniques at ADTPS, wherein imminent motor foundation damage of critical auxiliary DC Seal Oil Pump 1A was diagnosed.

Vibration monitoring and analysis helped in diagnosing the exact root cause of abnormity of motor foundation problem at very early stage and thus avoided the catastrophic failure of generator. This practice has saved the plant from huge losses in terms of generation, unplanned breakdown time and associated maintenance cost.

On 7 July 2013, during routine monitoring, it was observed that horizontal vibrations of motor DE and NDE bearings were deviated from 5.2 to 7.1 and 5.4 to 7.6 mm/s (RMS), respectively, even though the pump vibration was found normal. Then, it was decided to keep a close watch on vibration readings of motor and pump during daily trial.

On 8 July 2013, horizontal vibration reading of motor DE was increased from 7.1 to 11.6 mm/s. Vibration spectrum analysis revealed that there is increase in 1× frequency. On observing increasing trend of horizontal vibration of motor, decision was taken to carry out the necessary action. Increase in 1× frequency only in horizontal direction of motor can be due to misalignment between motor and pump or foundation problem of the motor. To confirm this exact root cause of abnormality, readings were taken on motor and pump foots which show uneven readings on motor foot that confirmed foundation problem of motor. With this conclusion, motor foundation rectification was carried out by providing additional adjustable supports to all motor foots ensuring all foots are in one plane means the level of motor foundation and pump foundation came one line which can not affect the alignment and foundation mismatch during pump running condition.

22.1 PRE-MAINTENANCE HISTORY

As shown in Figures 22.1 and 22.2, on 7 July 2013 during routine monitoring, it was observed that horizontal vibration of motor DE and NDE bearing was deviated from 5.2 to 7.1 and 5.4 to 7.6 mm/s (RMS), respectively even though pump vibration was found normal.

All other pump parameters (i.e., pressure, temperature) were found normal. It was decided to keep a close watch on vibration readings of motor and pump during daily trial. On 8 July 2013, horizontal vibration reading of motor DE was increased from 7.1 to 11.6 mm/s. By observing such increment trend of horizontal vibration, it was decided to carry out necessary action.

 DOI: 10.1201/9781003432616-22

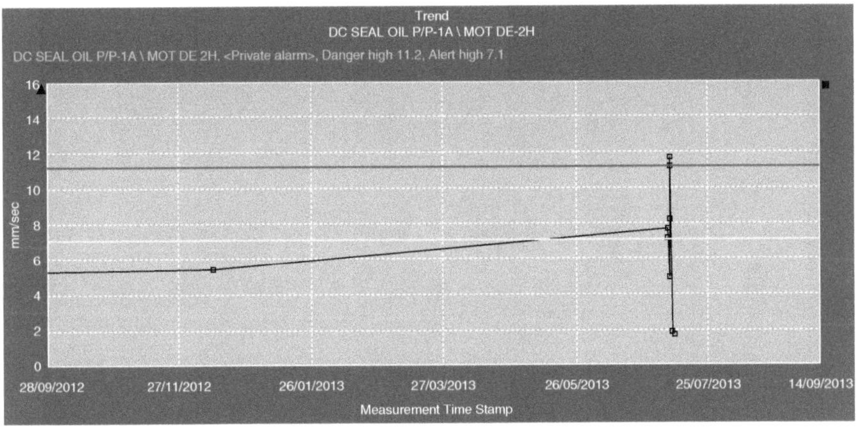

FIGURE 22.1 Vibration trend of motor DE (horizontal).

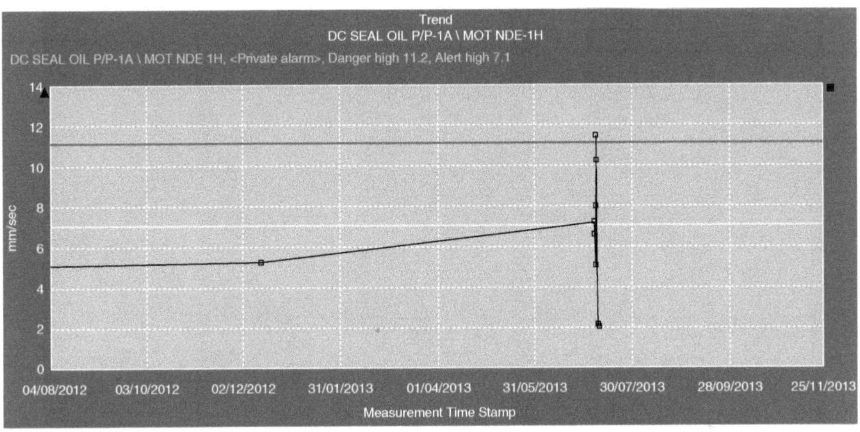

FIGURE 22.2 Vibration trend of motor NDE (horizontal).

22.2 VIBRATION DIAGNOSIS

On 8 July 2013, horizontal vibrations reading of DE and NDE motor were increased to 11.6 and 11.4 mm/s, respectively (same readings remain normally in the range of 4.0–5.0 mm/s). The spectrum analysis for root cause was investigated as follows:

22.3 FREQUENCY ANALYSIS

It was observed that amplitude of horizontal vibration of motor DE and NDE bearing were gradually increased from 5.2 to 7.1 mm/s and from 7.1 to 11.6 mm/s, respectively. Earlier it was found in the range of 4.0–5.0 mm/s. From Figures 22.3 and 22.4, spectrum analysis revealed that there was increase in 1× frequency. Bearing acceleration readings were found within limit which eliminated the possibility of existence of any bearing defect. Such increment in 1× frequency only in horizontal direction

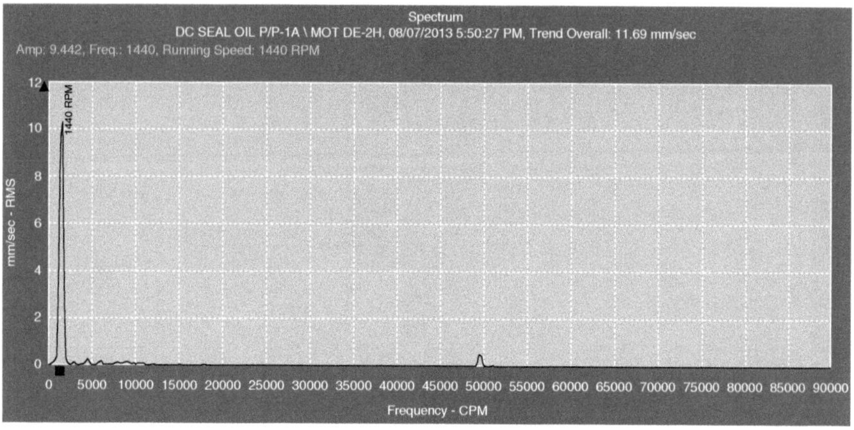

FIGURE 22.3 Vibration spectrum of motor DE (horizontal).

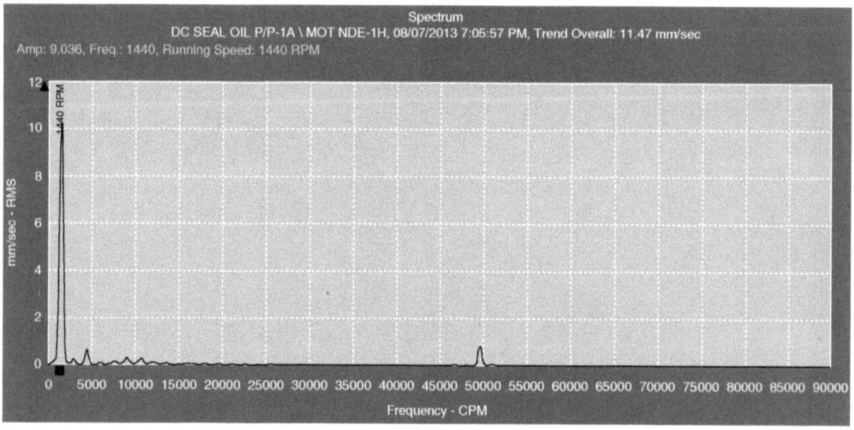

FIGURE 22.4 Vibration spectrum of motor NDE (horizontal).

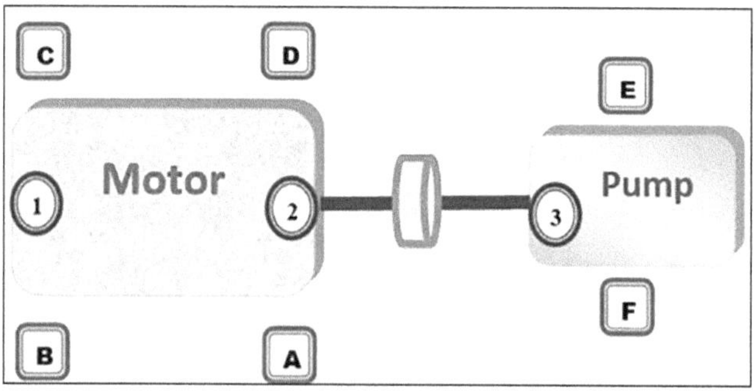

FIGURE 22.5 Offline vibration data points of motor and pump foot (at coupled condition).

can be due to misalignment between motor and pump or foundation problem of the motor. To confirm the exact root cause of abnormality, more readings were taken on all motor and pump foots as shown in Figure 22.5.

After analyzing all collected database, it was observed that pump foot readings were found in similar range, but motor foot readings were not stable and observed that all four foots were not found in one plane (refer Table 22.1).

The above analysis eliminated the possibility of misalignment between motor and pump because no deviation was noticed for pump DE bearing. To confirm the existence of motor foundation problem, de-coupled motor trial was taken.

As per the condition mentioned in Figure 22.6, only horizontal vibrations of motor DE and NDE bearing were found on higher side with predominant 1 × vibration frequency. Also, all motor foot readings were showing similar uneven pattern, which was observed in coupled condition (Table 22.2).

The above analysis confirmed that uneven foundation problem of motor was creating high horizontal vibration to the motor in decoupled condition and no influence of pump was observed. Decision was taken to rectify motor foundation and re-align with rest of the drive.

TABLE 22.1
Motor and Pump Foot Offline Vibration Data Comparison

Point Position (as Shown in Fig. 2.21.5)	Overall Vibration Velocity (mm/s) RMS (Vertical)
A	3.3
B	4.8
C	6.5
D	7.5
E	1.8
F	1.9

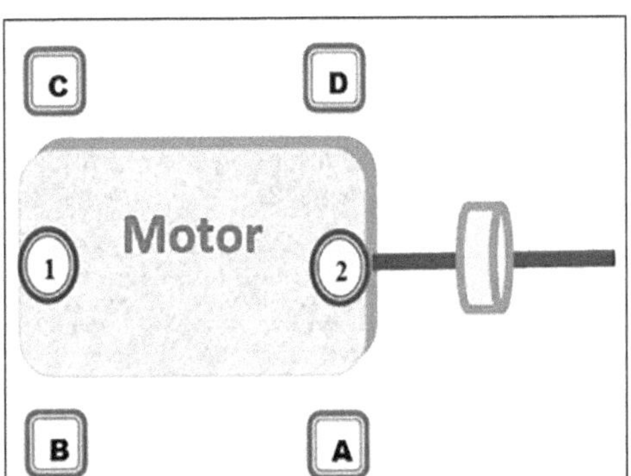

FIGURE 22.6 Motor foot offline vibration data (at decoupled condition).

TABLE 22.2
Motor Bearing and Foot Offline Vibration Data Comparison

Point Position (as Shown in Fig. 2.21.6)	Overall Vibration Velocity (mm/s) RMS
1 – H	5.1
2 – H	5.2
A – V	2.2
B – V	3.1
C – V	4.3
D – V	5.2

22.4 OBSERVATIONS RECORDED DURING MAINTENANCE WORK

Alignment between motor and pump was checked as shown in Figure 22.7. All readings were found within limits. Condition of coupling spider (rubber element of coupling) was found good. In addition, condition of motor and pump foundation bolts was found okay. Pump bearing inspection was done and bearing clearance recorded within limit (refer to Figure 22.8).

FIGURE 22.7 Alignment check between motor and pump.

FIGURE 22.8 Pump bearing clearance check.

22.5 MAINTENANCE WORK PERFORMED

Motor was de-coupled from pump. Before dismantling, initial alignment readings were taken for reference. Motor foundation rectification was done by providing additional adjustable supports to all motor foots ensuring that all foots are in one plane (Figure 22.9).

Alignment with pump was carried out and trial was taken. Vibrations of motor and pump bearing were found within limit. Also, readings were taken on motor and pump foot and observed that all readings were found in the same range (refer Table 22.4). Pump boxed up and made it available for operation.

22.6 POST-MAINTENANCE OFFLINE VIBRATION READINGS

Tables 22.3 and 22.4 represent comparison of vibration data of motor bearing and foundation foot for pre- and post-maintenance at full load condition. All values were reduced to normal level after providing additional support to motor foot. This was concluded with the success of work executed by turbine maintenance team.

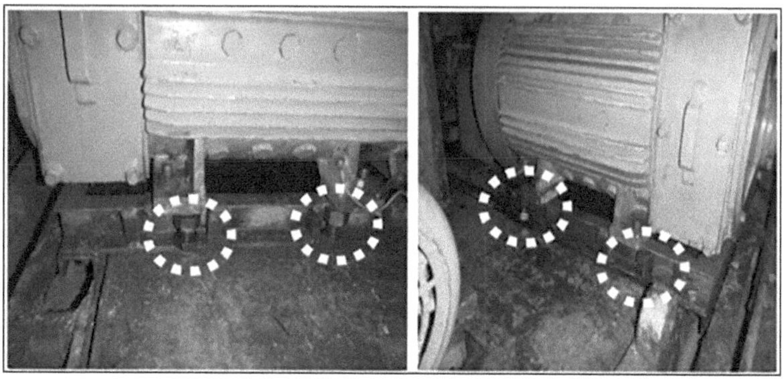

FIGURE 22.9 Additional supports provided to all motor foot.

TABLE 22.3
Pre- and Post-Maintenance Vibration Data

EQUIPMENT : DC SOP 1A DATE : 09 July 2013

LOCATION: - TG BASEMENT Unit - 01

VIBRATION REPORT OF DC SEAL OIL PUMP – 1A

Measurements			Overall Vibration Levels			
			Initial Readings 08/07/2013		After Rectification in Motor Foundation 09/07/2013	
Point		Position Date ->	Velocity (mm/s) RMS	Acceleration (g)	Velocity (mm/s) RMS	Acceleration (g)
Motor	(NDE)	H	**11.4**	0.8	2.1	0.7
		V	4.2	-	2.0	-
		A	1.6	-	0.9	-
Motor	(DE)	H	**11.6**	0.6	**1.8**	0.5
		V	4.3	-	1.7	-
		A	0.9	-	0.4	-
Pump	(DE)	H	0.8	0.2	1.0	0.2
		V	1.1	-	1.2	-
		A	1.1	-	1.0	-

22.7 POST-MAINTENANCE FOUNDATION VIBRATION READINGS

TABLE 22.4
Pre- and Post-Maintenance Motor and Pump Foot Offline Vibration Data

Point Position (as per Fig. 2.21.5)	Initial Readings Velocity (mm/s) RMS (Vertical)	After Rectification in Motor Foundation Velocity (mm/s) RMS (Vertical)
A	3.3	1.5
B	4.8	1.4
C	6.5	1.6
D	7.5	1.5
E	1.8	1.2
F	1.9	1.3

22.8 CONCLUSIONS

Vibration monitoring and analysis helped in diagnosing the exact root cause of abnormity of motor foundation problem at very early stage and thus avoided the catastrophic failure of generator. This has saved the plant from huge losses in terms of generation, unplanned breakdown time and associated maintenance costs.

DC powered seal oil pump supplies oil to the shaft seals of generator. Creating a pressurized oil film (acts as a seal) inside the generator which is higher than hydrogen pressure. Thus, it prevents escaping of hydrogen to the atmosphere which may lead to explosive mixture.

In the event of AC supply failure, DC seal oil pump automatically starts and takes over the oil supply to the shaft seals. Non-availability of DC seal oil pump during maintenance work will lead to escaping of hydrogen to the atmosphere and create damage to generator may cause for catastrophic failure.

23 Case Study 22
Gas Turbine 01 (BSES Kerala Plant): Misalignment of Generator – October 2012

EXECUTIVE SUMMARY

This is the success story of the implementation of condition monitoring techniques at BSES Kerala Power (BKPL) Kochi, wherein imminent misalignment between reducer gearbox and Gas Turbine Generator (GTG)-01 was diagnosed by providing expert service of ADTPS condition monitoring analyst.

Vibration and phase analysis helped in diagnosing the exact root cause of abnormity of misalignment between reducer gearbox and generator at very early stage and thus avoided the large catastrophic failure. This has saved the plant from huge losses in terms of generation, unplanned breakdown time and associated maintenance costs.

On 18 September 2012, GTG-01 was started after being in backed down condition (operated at low capacity than rated) for more than 18 months. Gas turbine-1 (GT-1) was running at about 16 MW on Naphtha, suddenly there was a load excursion, load of 16 MW dropped to zero MW with generator breaker in closed condition. After some time, machine was restarted and found generator voltage has built up to only 7 kV (rated is 11 kV) after exciting the machine and found that generator shaft was not rotating indicating a shear tube failure in the safeset in the load gearbox.

On 19 September 2012, Gas Turbine (GT) was started after safeset shear tube replacement. As the machine was loaded, gear box vibrations showed an increasing trend and moved toward trip limits. Hence, GT was stopped. On 20 September 2012, as per OEM instructions, gearbox internal inspection was carried out, and no abnormality was noticed.

On 21 September 2012, as advised by GE Engineering (OEM), safeset was again released, and machine was restated. At 10 MW load condition, vibration level was found on higher side and hence machine was withdrawn. On 22 September 2012, machine was started again for data collection by ADTPS condition monitoring analyst.

On 23 September 2012, GT-1 tripped due to NetCon controls card failure and during hot start, while ramping to Full Speed No Load (FSNL) condition, vibration values were found on higher side. Offline vibration and phase readings were taken. This matter was further taken up with original equipment manufacturer and has suggested for trying to correct the misalignment (recommendation by ADTPS analyst) by performing a dynamic balancing.

23.1 TECHNICAL DETAILS AND SCHEMATIC OF GTG-01

LM6000PC aero derivative gas turbine rated for 60 Hz application is connected to the 50 Hz generator through a reduction gear box. The low-speed shaft of the load coupling is connected to the generator through a torque limiting safeset (hydraulic type) coupling (Figures 23.1 and 23.2).

DOI: 10.1201/9781003432616-23

FIGURE 23.1 Aero derivative gas turbine (LHS) and reducer gearbox (RHS).

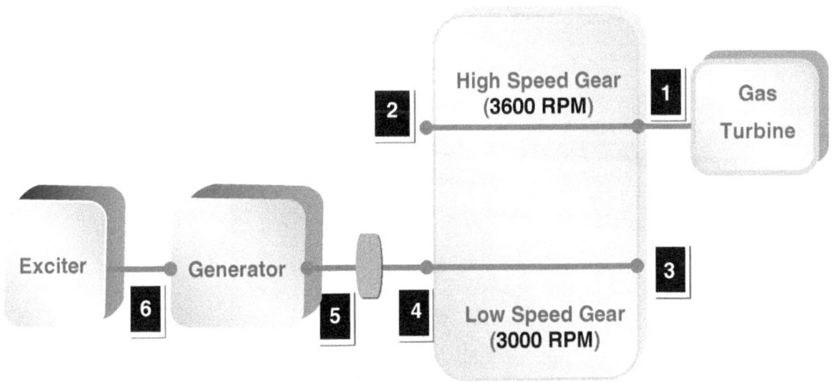

FIGURE 23.2 Gas Turbine Generator (GTG) power transmission schematic.

23.2 PRE-MAINTENANCE HISTORY

On 18 September 2012, plant was started after being in backed down condition for more than 18 months. Gas turbine-1 was running at about 16 MW on Naphtha. Suddenly, there was a load excursion, the load (MW) value dropped to zero along with generator breaker remained in closed condition (means generator connected to grid). Then after, generator circuit breaker was opened manually. Low pressure (LP) turbine speed went up to 3980 rpm turbine from the running speed of 3627 rpm. The gas turbine was hand-tripped. Machine was restarted and generator voltage built up to only 7 kV (actually rated is 11 kV). After exciting (synchronization) the machine, it was found that generator shaft was not rotating which indicate a shear tube failure in the safeset coupling of load gearbox (Figure 23.3). On 19 September 2012, the gas turbine was started after replacement of safeset shear tube and re-pressurizing the oil as per calibration chart provided by OEM M/s GE Limited (refer to Figure 23.4). Once the machine was loaded, online vibration readings of load gear box showed an increasing trend and moved toward trip limits. Hence, GT was stopped. On 20 September 2012, with reference to suggestion received from GE Engineering (i.e., OEM), some checks were done for reducer gearbox as shown in Figures 23.5 and 23.6. No abnormality/deviation was noticed. On 21 September 2012, after inspection

FIGURE 23.3 Shear tube location at safeset coupling of load gearbox.

FIGURE 23.4 Safeset coupling oil charging port.

FIGURE 23.5 Clearances between torque shaft and quill shaft in the load gear box.

FIGURE 23.6 Load gear box quill shaft gear teeth visual check via inspection window.

	Warm up cycle	Turbine LP (Inch/Sec) (A/B)	Turbine HP (Inch/Sec) (A/B)	Gen DE End Mils (A/B)	Gen NDE End Mils (A/B)	Gear Box (Inch/Sec) (A/B)
22/09/2012	0:00	0.2/0.15	0.19/0.07	1.14/1.3	1.43/1.33	0.08/0.09
	0:00	0.32/0.1	0.21/0.07	1.14/1.29	1.40/1.33	0.08/0.09
	0:00	0.33/0.08	0.20/0.08	1.43/1.28	1.39/1.36	0.09/0.10
	0:00	0.33/0.07	0.27/0.12	1.44/1.28	1.39/1.37	0.10/0.11
	0:00	0.38/0.11	0.34/0.18	1.48/1.28	1.36/1.38	0.11/0.20
	0:00	0.37/0.19	0.32/0.26	1.52/1.35	1.36/1.36	0.11/0.11

FIGURE 23.7 Online vibration data of gas turbine generator recorded on 22 September 2012.

and suggestions received from GE Engineering (OEM), safeset was again released, and machine was restarted. At 10 MW load condition, vibration level was found on higher side, and hence machine was withdrawn (taken out from service). On 22 September 2012, machine was started again for data collection by condition monitoring analyst. Offline vibration and phase data were collected at 32 MW load condition.

All vibration levels (mentioned in Figures 23.7 and 23.8) at different speeds and load conditions on all bearings were found within limit. On 23 September 2012, GT-1 tripped due to failure of NetCon controls card. During hot start, online vibrations were found on higher side while ramping to Full Speed No Load (FSNL) condition.

Then after, offline vibration and phase readings were taken at FSNL condition (refer to Figure 23.9).

On 23 September 2012, high vibration issue was further taken up with original equipment manufacturer (OEM) and has suggested to correct the misalignment (which was recommendation by ADTPS analyst) by performing a dynamic balancing. From 12 to 13 October 2012, in the presence of OEM experts, in-situ balancing was tried. The vibration levels at different balancing weights and load conditions on

MEASUREMENTS		OVERALL VIBRATION LEVELS			
Point	Position	Velocity (mm/sec) RMS			
Bearing No. as per figure		200 rpm @No Load	3000 rpm @No Load	20 MW Load	45 MW Load
1	H	0.07	2.7	3.9	4.8
	V	0.16	3.0	2.4	2.0
	A	0.08	1.2	2.6	2.2
2	H	0.09	3.0	4.3	2.7
	V	0.09	1.2	1.6	2.0
	A	0.09	2.5	2.9	2.2
3	H	0.07	1.8	2.6	2.2
	V	0.05	1.0	1.5	1.8
	A	0.06	1.8	2.0	1.8
4	H	0.05	2.2	2.5	1.8
	V	0.05	1.6	1.9	1.8
	A	0.06	1.6	2.1	1.7
5	H	0.03	1.4	4.0	2.5
	V	0.05	1.3	3.5	2.7
	A	0.05	2.4	2.2	3.5
6	H	0.04	0.7	1.0	1.3
	V	0.04	0.5	1.1	0.6
	A	0.05	0.6	1.2	1.8

FIGURE 23.8 Offline vibration data of gas turbine generator recorded on 22 September 2012.

all bearings were not found stable that indicated non repeatability on higher side. It was decided to go for re-alignment of generator and reducer gearbox with rest of drive.

23.3 VIBRATION DIAGNOSIS

On 23 September 2012, horizontal vibrations of reducer gearbox bearing (at FSNL condition) were found on higher side. The spectrum analysis for root cause was investigated as follows:

23.4 FREQUENCY ANALYSIS

The vibration levels on all gearbox bearings at no load (3000 rpm condition) were found on higher side only in horizontal direction. As shown in Figure 23.10, predominant frequency was found at 3000 rpm and also it was found at 3600 rpm for gear shaft bearing (refer to Figure 23.2). Phase difference in axial direction across the coupling between gearbox and generator bearing (i.e., bearing 4 and 5) was found 128° (refer to Figure 23.11 for phase spectrum and Figure 23.2 for bearing location).

EQUIPMENT : GAS TURBINE GENERATOR 01		DATE : 23rd SEPTEMBER 2012	
LOCATION: - BSES KERALA POWER, KOCHI. VIBRATION REPORT OF GTG 01			
MEASUREMENTS		OVERALL VIBRATION LEVELS	
Point	Position		
Bearing No. as per figure		Vibration Data mm/sec (RMS) 3000 rpm @No Load	Phase Data Degrees 3000 rpm @No Load
1	H	12.6	220
	V	6.6	218
	A	6.8	14
2	H	15.1	258
	V	9.4	27
	A	5.7	166
3	H	7.7	330
	V	9.3	227
	A	4.3	128
4	H	10.3	223
	V	7.5	265
	A	3.6	234
5	H	2.2	221
	V	1.2	183
	A	2.3	106
6	H	0.5	96
	V	0.5	252
	A	0.3	86

FIGURE 23.9 Offline vibration and phase readings recorded on 23 September 2012.

Phase difference in axial direction across the coupling between gearbox and exciter bearing (i.e., bearing 4 and 6 as shown in Figure 23.2) was found 148°. Axial movement of both gearbox shafts was observed at generator and turbine side at 3000 rpm under the strobe light (shown in Figure 23.13). It is observed that only axial vibration spectrum of gearbox bearings was having unknown frequency at 2363 rpm (refer to Figure 23.14), which may be due to axial pull at the coupling between reducer gearbox and generator. From Figures 23.10 to 23.14, the vibration and phase analysis confirmed misalignment between gearbox and generator.

23.5 PERFORMED MAINTENANCE WORK MISALIGNMENT CORRECTION BY BALANCING

ADTPS condition monitoring analyst recommended to decouple the generator/gear box and do the alignment check. Since decoupling the generator and realigning

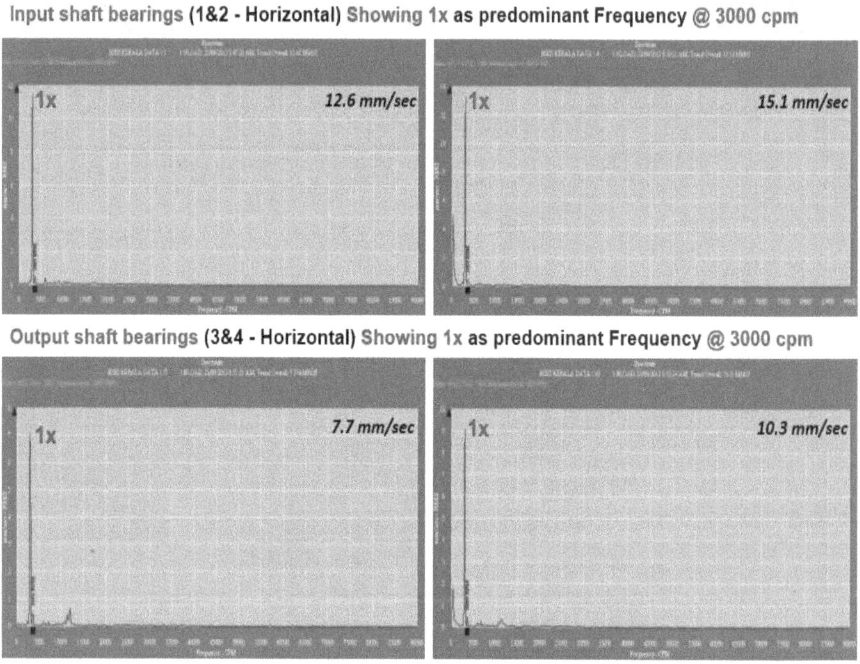

FIGURE 23.10 Vibration spectrum of gearbox input and output bearing (horizontal).

the shaft is a laborious exercise, GE Engineering (OEM) suggested to correct the misalignment by performing a dynamic balancing. On 12 October 2012, 0°–360° marked on torque shaft (generator shaft) with interval of 30° for addition of balancing weights (refer to Figure 23.15).

To review the symptoms of vibration due to unbalance, counterbalance weights were added up in the generator coupling. First 60 g of balancing weight was added in the slot which was provided in the generator coupling corresponding to 0° marking (as shown in Figure 23.16).

GT was started and ramped to FSNL condition. Vibrations (generator DE and gearbox) were found on higher side (shown in Figure 23.17). Hence, machine was not loaded and stopped. Balancing weight of 60 g was replaced with 27.8 g at 180° position. Vibration readings were found normal at FSNL condition as shown in Figure 23.18. On 13 October 2012, for verifying the repeatability of readings, balancing weight of 40 g was added at 180° after removing the 27.8 g. Trial was taken, and high vibration readings were observed at FSNL condition which indicated the non-repeatability (non-consistency) of readings (as shown in Figure 23.19). Machine was withdrawn, and 40 g counterweight was replaced with 27.8 g. GT was synchronized and loaded to 32 MW. As the machine picked up load, online vibration readings were increased simultaneously (refer to Figure 23.20). The vibration levels at different balancing weights and load conditions on all bearings were not found stable, which indicated non repeatability (on higher side). It was decided to go for re-alignment of generator and reducer gearbox.

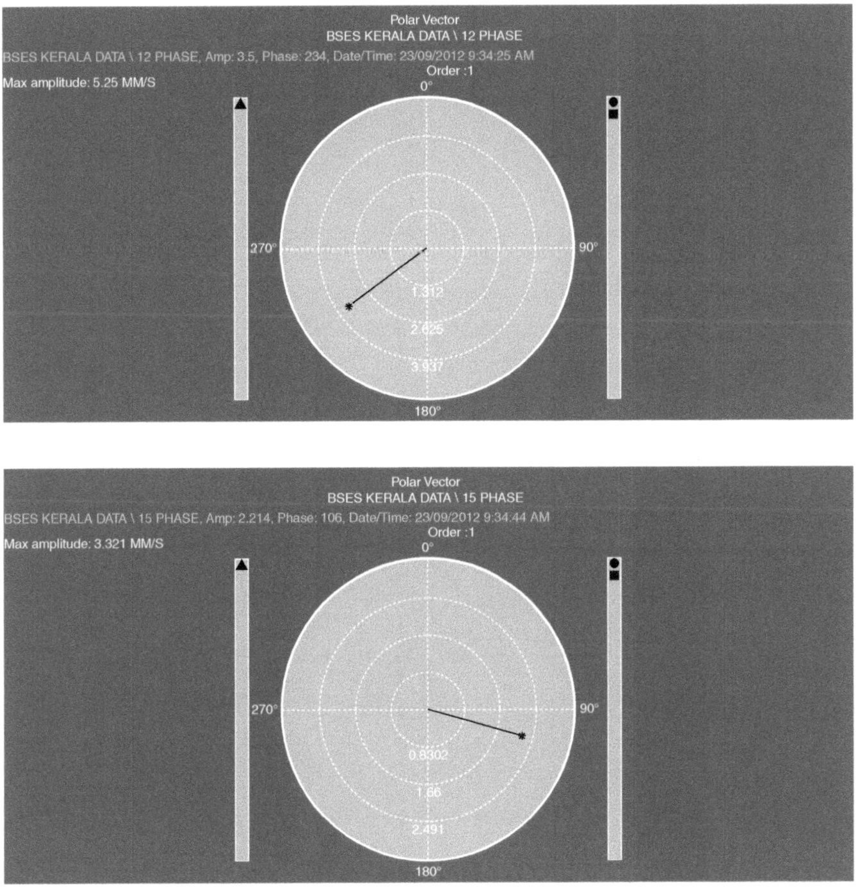

FIGURE 23.11 Phase difference in axial direction across the coupling between bearing 4 and 5.

23.6 PERFORMED MAINTENANCE WORK MISALIGNMENT CORRECTION BY DE-COUPLING

On 14 October 2012, generator front bearing thrust pad was removed as shown in Figure 23.22. On 16 October 2012, with the help of jack bolts, generator rotor was pushed toward exciter (axially 7.5 mm) as suggested in manual which was provided by GE Engineering (OEM). The pre-stretch on the load coupling was found approximately 4.3 mm. Since rear generator bolt (RGB) loses its axial fixed point after decoupling (RGB does not have thrust bearings) the shafts will move toward the GT after decoupling. Thereby a coupling gap of 11.5 mm was achieved enough (shown in Figure 23.24) to release the rabbet and rotate the torque shaft for alignment checking. The initial radial readings were crossing the tolerance limits by 0.2 mm (refer to Figure 23.25). On 17 October 2012, old shims of generator were removed, cleaned and placed back along with the shims for correction (0.25 mm) for all eight holding bolts (shown in Figures 23.26 and 23.27). On 18 October 2012, alignment of

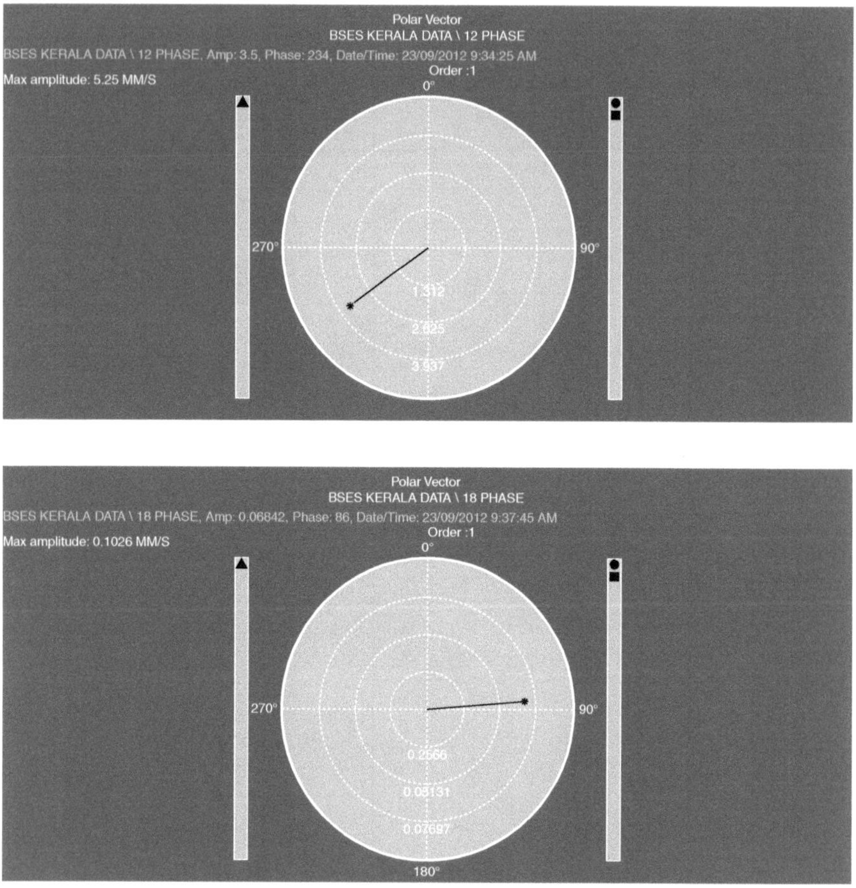

FIGURE 23.12 Phase difference in axial direction across the coupling between bearing 4 and 6.

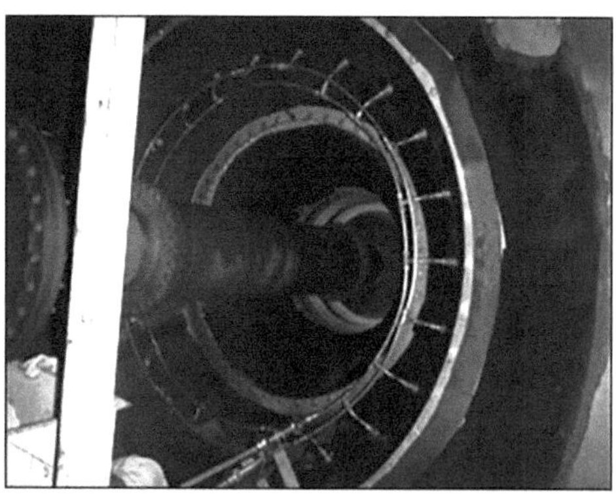

FIGURE 23.13 Axial shaft movement noticed under strobe light.

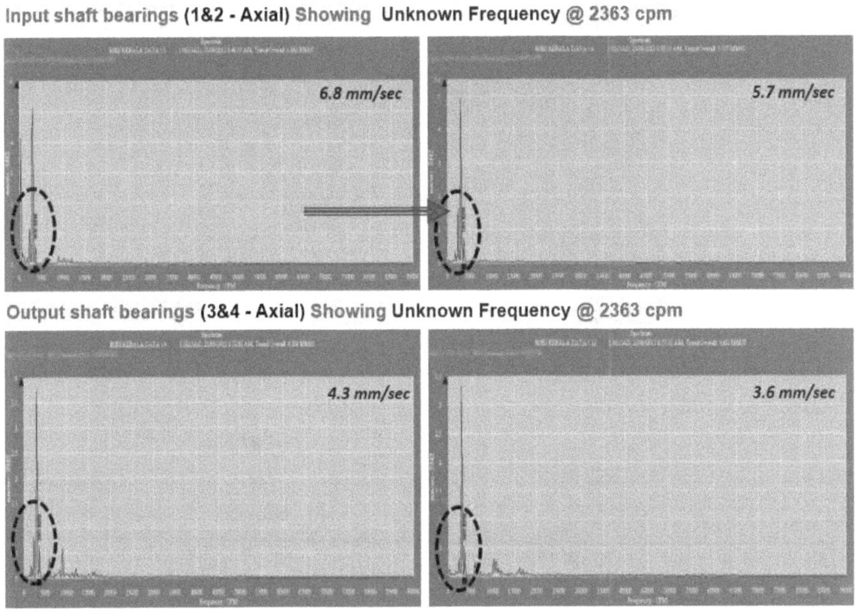

FIGURE 23.14 GB input and output bearing spectrum showing unknown frequency at 2363 rpm.

FIGURE 23.15 Reference marking done on generator shaft.

generator shaft was done with reference to low-speed torque shaft (generator shaft) of load gearbox after centering the torque shaft. Shear ring-safeset returned to normal position and pressurized. The alignment of radial readings was varied with increase in lube oil temperature as lube oil pump was continuously ON while performing the alignment (refer to Figure 23.28). On 19 October 2012, ultrasonic and liquid penetration test (LPT) were carried out for generator front journal bearing and its thrust pads. No abnormality was noticed (refer to Figure 23.29). On 20 October 2012, final alignment readings were taken, and all readings were found within prescribed limits (Figure 23.30).

FIGURE 23.16 Addition of trial weight (60 g) on generator shaft.

Date	Time	MW	Turbine LP (Inch/Sec) (A/B)	Turbine HP (Inch/Sec) (A/B)	Gen DE End Mils (A/B)	Gen NDE End Mils (A/B)	Gear Box (Inch/Sec) (A/B)
12/10/2012	17:15	During ramp up	0.22/008	0.14/0.00	0.92/0.98	1.93/1.17	0.44/0.42
		During warm up	0.27/0.7	0.14/0.00	1.21/1.06	1.69/1.6	0.44/0.4
		During warm up	0.43/0.57	0.1/0.0	1.07/0.87	1.6/1.52	0.45/0.39
		FSNL	0.37/0.6	0.18/0	1.01/0.8	1.53/1.46	0.46/0.4

FIGURE 23.17 Online vibration data after addition of trial weight (60 g).

Date	Time	MW	Turbine LP (Inch/Sec) (A/B)	Turbine HP (Inch/Sec) (A/B)	Gen DE End Mils (A/B)	Gen NDE End Mils (A/B)	Gear Box (Inch/Sec) (A/B)
12/10/2012	18:05	Core Idle	0.03/0.01	0/0	0.63/0.58	1.61/1.39	0.02/0
		Acc to Sync idle	0.06/0.05	0.05/0	0.63/0.7	1.56/1.4	0.03/0.03
		Warm up cycle	0.22/0.65	0.13/0	1.54/1.51	1.6/1.57	0.11/0.08
		Warm up cycle	0.41/0.6	0.13/0	1.38/1.31	1.47/1.48	0.1/0.08
		FSNL	0.26/0.72	0.08/0	1.35/1.26	1.44/1.45	0.01/0.07
		Cool down cycle	0.05/0.02	0.06/0	0.57/0.59	1.51/1.35	0.02/0.01

FIGURE 23.18 Online vibration data after addition of weight 27.8 g.

Date	Time	MW	Turbine LP (Inch/Sec) (A/B)	Turbine HP (Inch/Sec) (A/B)	Gen DE End Mils (A/B)	Gen NDE End Mils (A/B)	Gear Box (Inch/Sec) (A/B)
13/10/2012	14:45	Acc to core idle	0.03/0.01	0/0	0.65/0.64	1.53/1.34	0.02/0.01
		core idle	0.03/0.01	0.03/0	0.77/0.66	1.5/1.33	0.02/0.01
		Acc to Sync idle	0.32/0.14	0.11/0	1.25/1.57	1.53/1.53	0.16/0.23
		Warm up cycle	0.22/0.5	0.13/0	1.32/1.28	1.52/1.5	0.17/0.24
		Warm up cycle	0.48/0.5	0.08/0.0	1.23/1.15	1.49/1.46	0.18/0.24
		FSNL	0.46/0.58	0.15/0	1.18/1.11	1.48/1.45	0.18/0.25

FIGURE 23.19 Online vibration data after addition of weight 40 g.

Date	Time	MW	Turbine LP (Inch/Sec) (A/B)	Turbine HP (Inch/Sec) (A/B)	Gen DE End Mils (A/B)	Gen NDE End Mils (A/B)	Gear Box (Inch/Sec) (A/B)
13/10/2012	15:55	Warm up cycle	0.12/0.43	0.72/0	1.55/1.43	1.48/1.4	0.11/0.14
		Warm up cycle	0.3/0.66	0.06/0	1.44/1.31	1.41/1.39	0.11/0.13
		FSNL	0.37/0.58	0.15/0	1.38/1.23	1.4/1.4	0.12/0.14
		FSNL	0.25/0.59	0.16/0	1.48/1.2	1.45/1.36	0.21/0.23
		7.3MW	0.21/0.48	0.16/0	1.48/1.21	1.45/1.34	0.25/0.27
		8MW	0.26/0.15	0.17/0	1.44/1.24	1.46/1.32	0.27/0.28
		12MW	0.14/0.36	0.11/0	1.39/1.23	1.45/1.32	0.32/0.32
		17MW	0.16/0.33	0.11/0	1.35/1.23	1.43/1.34	0.39/0.37
		20MW	0.17/0.11	0.23/0	1.33/1.24	1.42/1.34	0.41/0.38
		25MW	0.26/0.1	0.27/0	1.33/1.29	1.4/1.34	0.42/0.41
		32MW	0.32/0.09	0.28/0	1.4/1.33	1.37/1.36	0.43/0.39

FIGURE 23.20 Online vibration data after addition of weight 27.8 g loaded up to 32 MW.

On 21 October 2012, full load trial of the machine was taken, and the vibration readings were found to be comparable with the readings prior to safeset coupling failure reported on 18 September 2012. Hot start of the machine was also performed. Vibrations were found to be well within limits during hot start up as well. The vibration levels at different speeds and load conditions on all bearings were found within limit.

23.7 POST-MAINTENANCE VIBRATION READINGS

On 12 October 2012, in the presence of OEM expert, in-situ balancing was tried. The vibration levels at different balancing weights and load conditions on all bearings were found not stable, also indicating non repeatability (not getting the same value each time). It was decided to go for re-alignment of generator and reducer gearbox.

Measuring point		Cold start (No counter weight(	Hot start (No counter weight)	Added wt of 28gm @ 180 deg at FSNL	Added wt of 40gm @ 180 deg at FSNL	Added wt of 28gm @ 180 deg at FSNL	Added wt of 28gm @ 180 deg and @ 5MW	Added wt of 28gm @ 180 deg and @ 12MW	Added wt of 28gm @ 180 deg and @ 25MW
		11:30	14:00	18:00					
BRG 1	H	7.3		3.1	12.9	6.9		6.9	7.1
	V	4.5		2.1	3.6	2.3			8.4
	A	3.6		2.3	6.4	2.4			3.9
BRG 2	H	7.5		3.6	14.8	6.9		7.5	7.5
	V	4.9		1.6	19.6	3			7
	A	3.8		2.3	6.5	3.5			3.5
BRG 3	H	4.8	4.8	1.9	8.2	4.2	4.3	4	6.9
	V	2.3	4.3	2.3	6	3.4			9
	A	2.7		1.7	4.9	1.9			2.4
BRG 4	H	5.1	6.1	2.6	10.3	4.9	5.3	5.1	5.2
	V	3.5	4.6	2	7.3	3.9			8.2
	A	2.4		1.5	4.3	1.7			2.2
BRG 5	H	2.1	1.2	1.4	1.5	2.1	2.6	2.9	2.5
	V	1.3	1	1.2	0.7	1.4			2.1
	A	2.3		2.2	1.8	2.5			2.6
BRG 6	H	0.7		0.7	0.7	0.6		0.9	0.8
	V	0.5		0.5	0.4	0.4			0.5
	A	0.8		0.7	0.6	0.7			1.6

FIGURE 23.21 Offline vibration data comparison with different balancing weights.

FIGURE 23.22 Removal of generator front bearing and its thrust pad.

FIGURE 23.23 De-coupling of generator from reducer gearbox.

FIGURE 23.24 Generator rotor shifted toward exciter for alignment work.

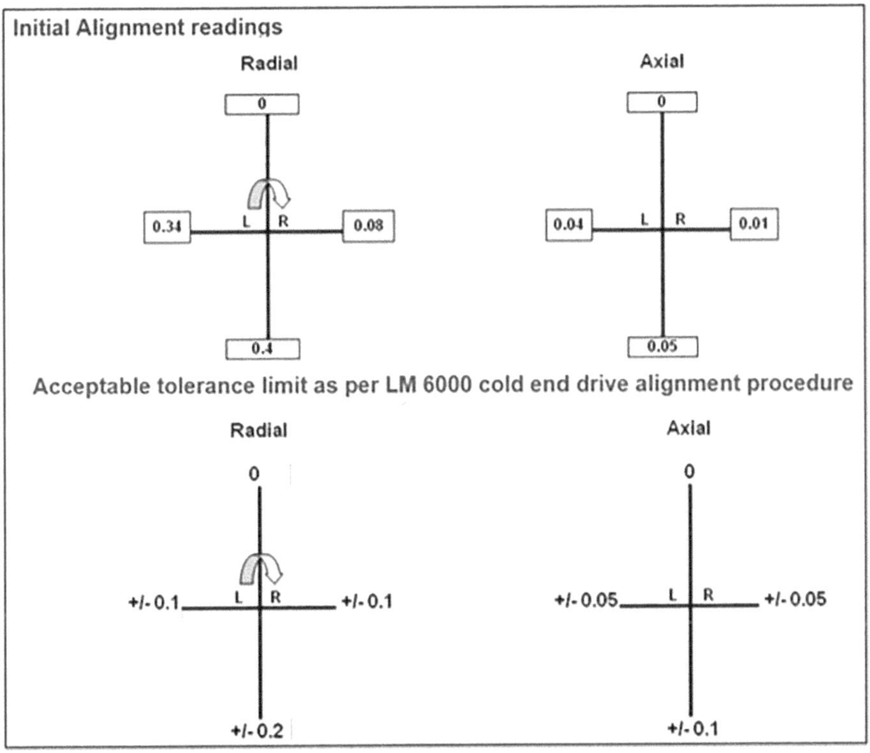

FIGURE 23.25 Initial alignment readings.

FIGURE 23.26 Removal of generator foot shims.

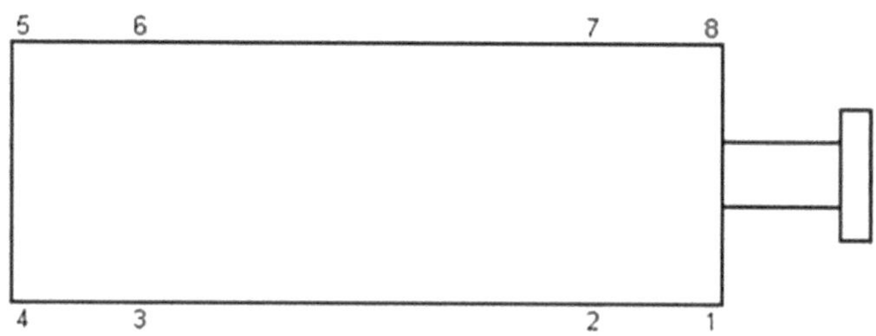

Bed bolt No:	1	2	3	4	5	6	7	8
Shim thickness	2.70 mm	2.70 mm	2.75mm	2.75mm	2.55mm	2.55mm	2.40mm	2.40mm

FIGURE 23.27 Alignment correction done at generator bolts.

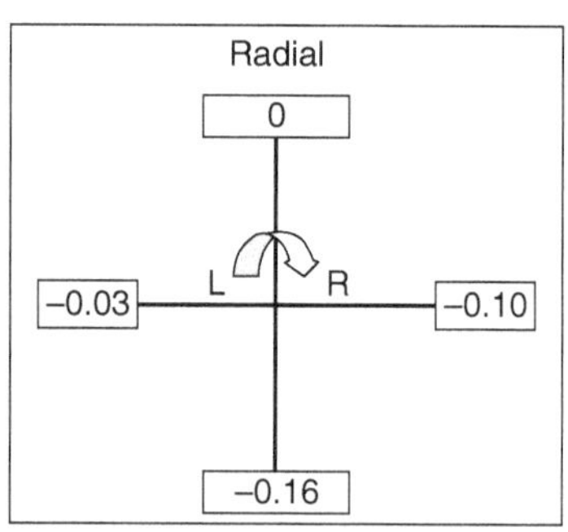

FIGURE 23.28 Alignment reading taken on 18 October 2012.

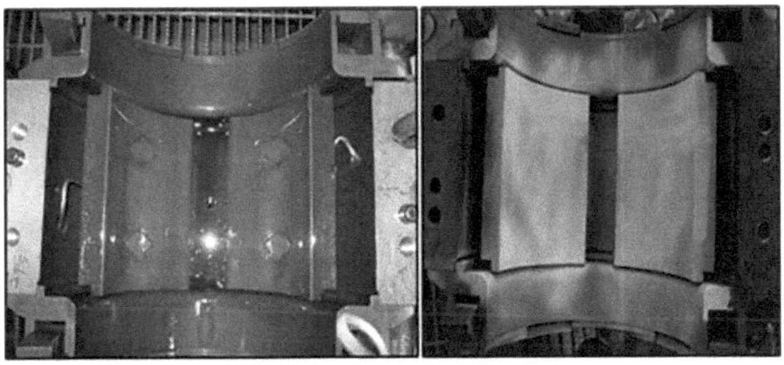

FIGURE 23.29 LPT test done for generator front journal bearing and its thrust pads.

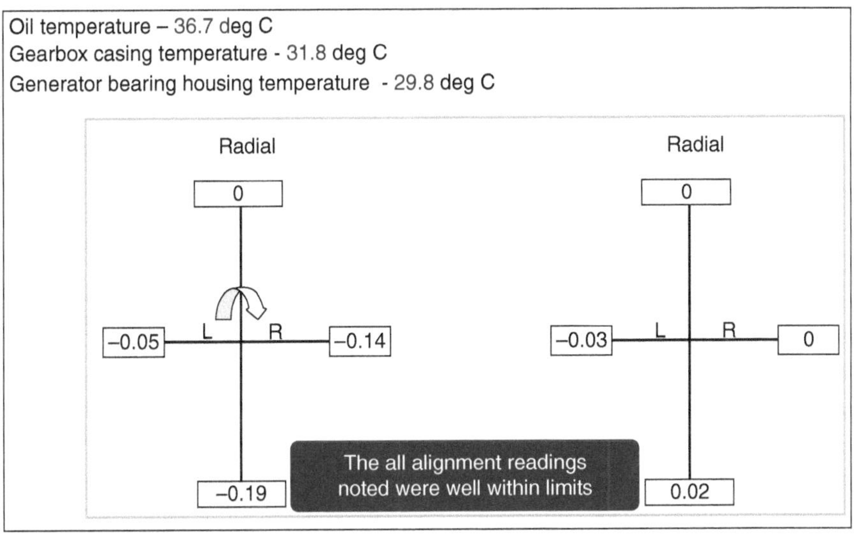

FIGURE 23.30 Final alignment reading.

On 14 October 2012, 0.25 mm shims added to all feet of generator and alignment work was completed on 20 October 2012. On 21 October 2012, the vibration levels at different speeds and load conditions on all bearings were found within satisfactory limit.

Figure 23.32 represents the comparison of online vibration data for pre- and post-maintenance. From Figure 23.31, all values were reduced to normal level after re-alignment work. It was concluded that maintenance work was executed successfully.

BSES KERALA POWER LIMITED

VIBRATION LIMITS

The vibration limits as specified by OEM for the equipments are as given below

Measuring point	Alarm	Trip
Turbine LP (Inch/Sec) (A/B)	1.4	2
Turbine HP (Inch/Sec) (A/B)	2	2
Gen DE End Mils(A/B)	3	4
Gen NDE End Mils (A/B)	3	4
Gear Box (Inch/Sec) (A/B)	0.3	0.5

FIGURE 23.31 Vibration alarm and trip reference values for GTG.

Date	Time	MW	Turbine LP (Inch/Sec) (Channel A/B)	Turbine HP (Inch/Sec) (Channel A/B)	Gen DE End (Mils Peak to Peak) (Channel A/B)	Gen NDE End (Mils Peak to Peak) (A/B)	Gear Box (Inch/Sec) (Channel A/B)	Remarks
13-Apr-11	21:00	424	0.40/0.25	0.27/0.16	0.42/0.40	1.42/1.42	0.08/0.10	2011 reading
18-Sep-12	18:35	16	0.28	0.09	0.43/0.36	1.45/1.43	0.06/0.08	Prior to machine trip and release of load gear box safe set coupling. Machine tripped at 16 MW and safe set pin got sheared
20-Sep-12	0:00	43.6	0.41/0.18	0.24/0.23	1.64/1.61	1.34/1.35	0.18/0.24	After resetting and repressurising the safeset
23-Sep-12	9:25	FSNL	0.15/0.64	0.13/0.2	1.15/0.8	1.46/1.32	0.41/0.39	While attempting a hot start post turbine trip
21-Oct-12	0:13	41.8	0.36/0.13	0.27/0.07	0.38/0.37	1.37/1.43	0.08/0.12	After centering the load gear box torque shaft and realigning the generator and gear box

FIGURE 23.32 Comparison of online vibration data for pre- and post-maintenance.

23.8 ROOT CAUSE ANALYSIS – FLOW CHART AND FISHBONE DIAGRAM

Figure 23.33 represents the detailed flowchart for resolving the high vibration issue experienced by Gas Turbine Generator (GTG), and Figure 23.34 discusses on various probable causes for such high vibration issues.

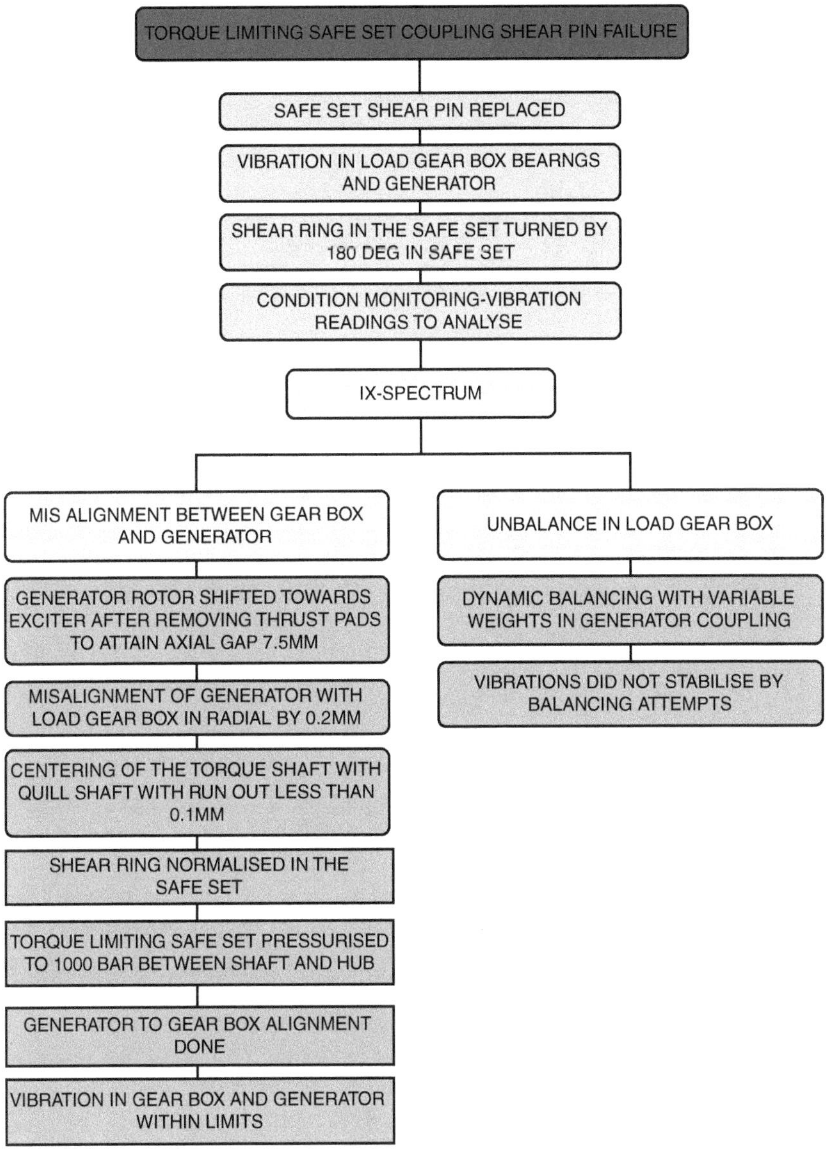

FIGURE 23.33 Flowchart.

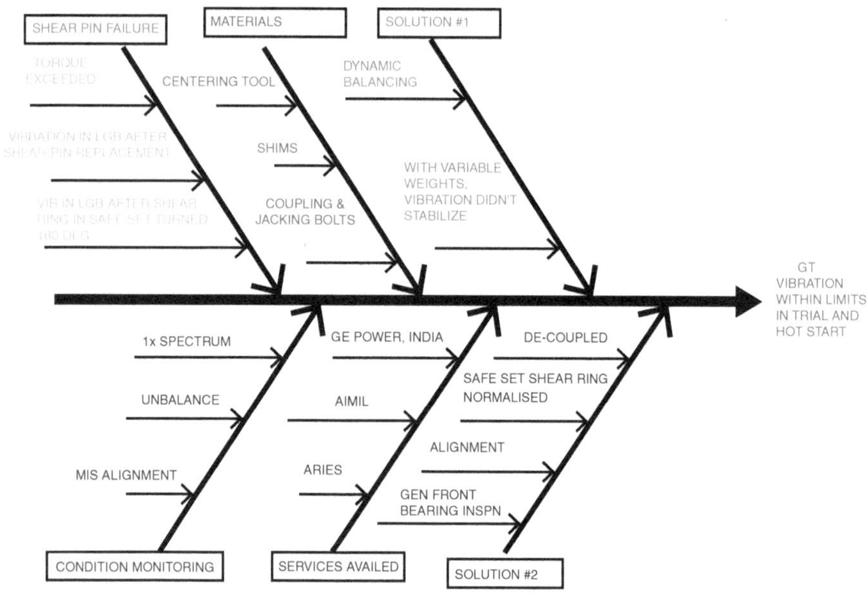

FIGURE 23.34 Fishbone diagram.

23.9 CONCLUSIONS

When the load gear box torque shaft and generator rotors were decoupled for verifying the alignment, it was found that the alignment values have crossed the limits (radial was 0.4 mm against the allowable limit of 0.2 mm). It was found that torque shaft run out within the low-speed gear (part no.3) shaft which found 0.6 mm (against limit of 0.1 mm). The torque shaft is supported by the safeset with its bearing on one side of the shaft. Other end of the toque shaft is supported by the generator drive end bearing in coupled condition. At this position, the vibration levels have been found satisfactory (there were no instances of high vibrations on this gear box since its commissioning until the safeset was released on 18 September 2012). However, when the safeset was released, the torque shaft has been running free. When re-pressurized, it will not be coupled on the exact same spot where it was originally coupled. The resulting unbalance has caused high vibrations of the load gear box.

No maintenance of the load gear box/generator was taken up since commissioning of unit which would have affected the alignment between load gear box quill shaft and generator rotor. While realigning the load gear box to the generator rotor, 0.25 mm thick shims added to all the four generator bed bolts. Based on alignment/run out as observed after decoupling the generator and load gear box, it can be reasonably concluded that the load gear box torque shaft was not centered during the initial alignment, and the generator was aligned to a torque shaft which was not properly centered.

The GTG 01 was not available due to high reducer gearbox vibrations and tripping at FSNL. After correcting misalignment between generator and reducer gearbox, GTG-01 was synchronized and at full load, all vibrations were found within limits and made available 100% for operation.

24 Case Study 23
ID Fan 1B: Motor Bearing Looseness – January 2011

EXECUTIVE SUMMARY

The present case study is a success story of the implementation of condition monitoring techniques at ADTPS, wherein imminent motor bearing failure due to looseness of bearing on shaft of HT auxiliary ID FAN – 1B was diagnosed.

Vibration analysis and noise monitoring helped in diagnosing the exact root cause of abnormity at very early stage. This has saved the plant from huge losses in terms of generation, unplanned breakdown time and associated maintenance cost, which could have caused total cost of approximately 2.15 crores.

ID fan was taken in service on 9 December 2010 after shutdown. On 22 December 2010, abnormal sound was noticed from NDE side of motor. On 5 January 2011, sound level was increased from 94 to 99.1 dB even though no deviation was observed in motor vibrations. In view of increasing trend of noise, it was decided to stop the motor for 2 hours to carry out internal inspection of motor cooling fan for any rubbing between stationary and rotating part. Inspection was carried out on the same day. Motor was taken in service, and no abnormality was observed.

On 6 January 2011, horizontal vibration reading of NDE and DE motors were deviated from 0.5 to 0.8 and 0.6 to 0.8 mm/s (RMS), respectively. This reading remained constant up to 9 January 2011. On 10 January 2011, noise level increased from 99.1 to 101.9 dB. Also, vibration of NDE motor was increased from 0.8 to 1.2 mm/s. In view of increasing trend of noise, it was decided to stop the ID fan for motor replacement. On 11 January 2011, motor was replaced with spare and noise, vibration and all other parameters were found well within satisfactory limits. Meanwhile during the inspection of motor, inner race of NDE side was found cracked and loosen over the shaft. In addition, scoring marks was observed over the shaft. No damage to rotor and stator windings were observed after thorough inspection.

24.1 PRE-MAINTENANCE HISTORY

On 25 November 2010, during unit overhaul, it was decided that ID fan 1B motor to be replaced with the serviced spare motor as per maintenance schedule. On 26 November 2010, no load decoupled trial was taken for 1 h wherein motor drive end (DE) and non-drive end (NDE) bearing vibrations were found within limit. At the same time, DE and NDE bearing temperatures were also found normal. On 9 December 2010, fan was taken in service. Vibration readings of motor, hydraulic coupling and fan were found within prescribed limit. Also, all the operating parameters were found normal. On 22 November 2010, suddenly abnormal sound was noticed from NDE side of motor. Immediately, vibration readings were taken, and

DOI: 10.1201/9781003432616-24

all readings were found well within limits. Noise level was measured for reference, which was found 94dB (refer to Figure 24.1). Such abnormal sound level remained constant for next 10days. But no variations in vibration readings were noticed. On 5 January 2011, noise was increased to 99.1dB, still no deviation was observed in motor vibration (refer to Figure 24.2). By observing such increasing trend of noise from NDE side of motor, it was decided to stop the motor for 2hours and carry out the internal inspection of motor cooling fan to ensure any rubbing between stationary and rotating part. Inspection was carried out. No abnormality has been observed. Motor was taken in service. On 6 January 2011, horizontal vibration readings of motor NDE and DE were deviated from 0.5 to 0.8 and 0.6 to 0.8 mm/s (RMS), respectively. At the same time, there was no vibration deviation found in hydraulic coupling and fan bearing. On 8 January 2011, detailed vibration analysis report along with various probable causes was discussed in HoD meeting. It was decided to keep the close watch on motor vibration and noise readings. As shown in Figures 24.3 and 24.4, on 9 January 2011, vertical vibration reading of motor NDE and DE deviated from 0.7 to 1.0 and 0.6 to 0.9 mm/s (RMS), respectively. On 10 January 2011, horizontal vibration of motor NDE and DE was increased to 1.3 and 1.2 mm/s and vertical vibration also increased to 1.1 and 1.3 mm/s, respectively. On 10 January

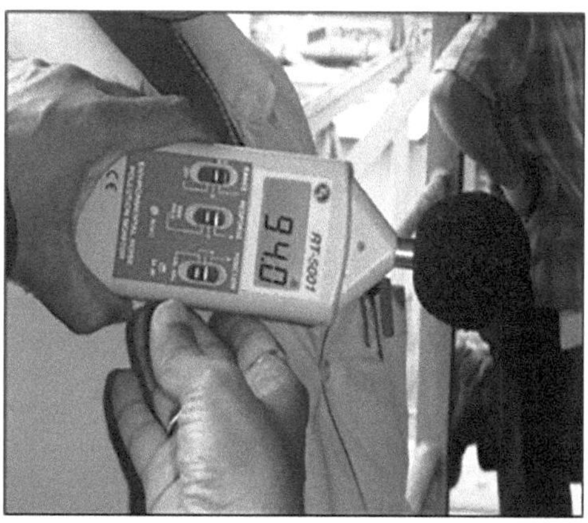

FIGURE 24.1 Noise data collection at motor NDE side on 22 November 2010.

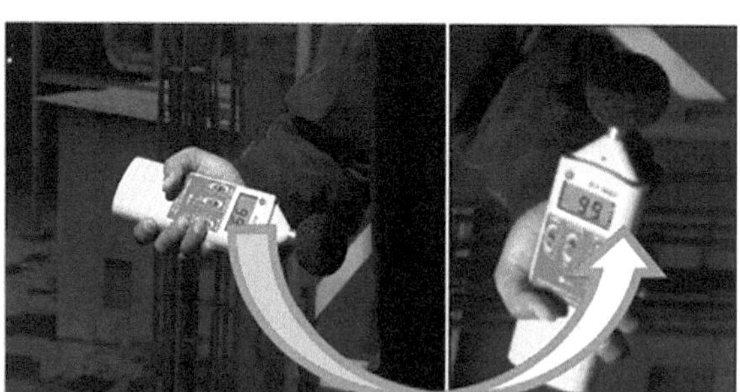

FIGURE 24.2 Noise data collection at motor NDE side on 5 January 2011.

2011, noise level further increased from 99.1 to 101.9 dB as reported in Figure 24.5. By observing such increase trend of noise from 94 to 101.9 dB, it was decided to stop the ID fan for motor replacement.

24.2 VIBRATION DIAGNOSIS

On 10 January 2011, horizontal vibration readings of NDE and DE motor were increased from 0.5 and 0.6 to 1.3 and 1.2 mm/s. Similarly, vertical vibration readings were increased from 0.7 and 0.6 to 1.1 and 1.3 mm/s (RMS), respectively. The spectrum analysis for root cause was investigated as follows:

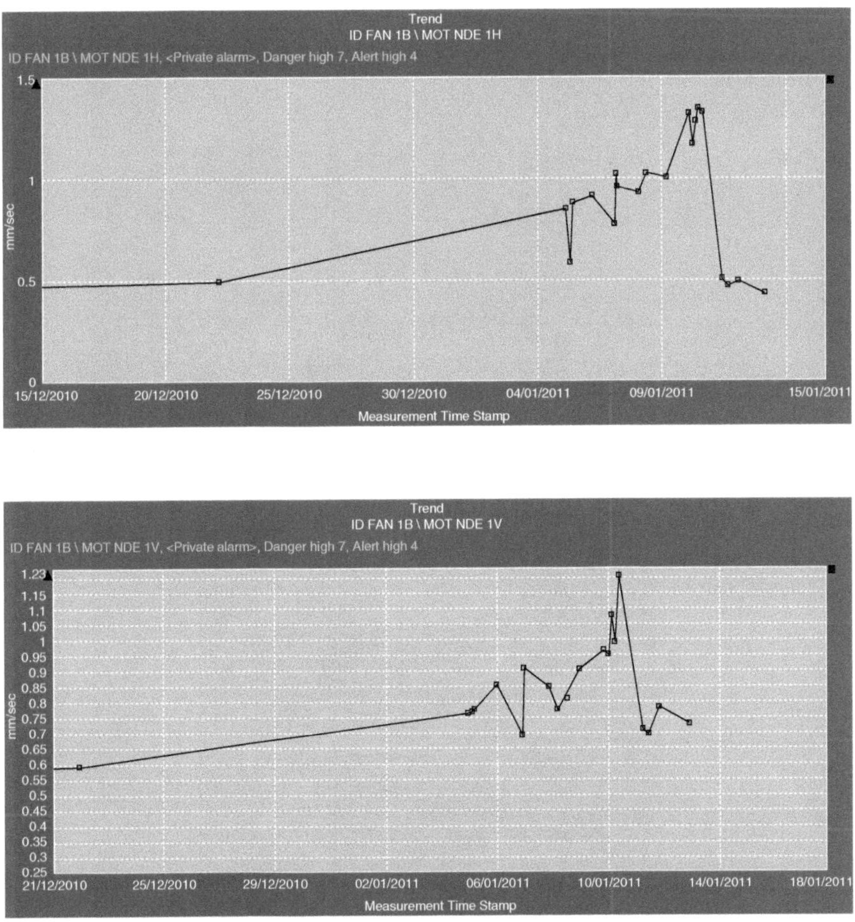

FIGURE 24.3 Vibration trend of ID Fan 1B motor NDE (horizontal and vertical).

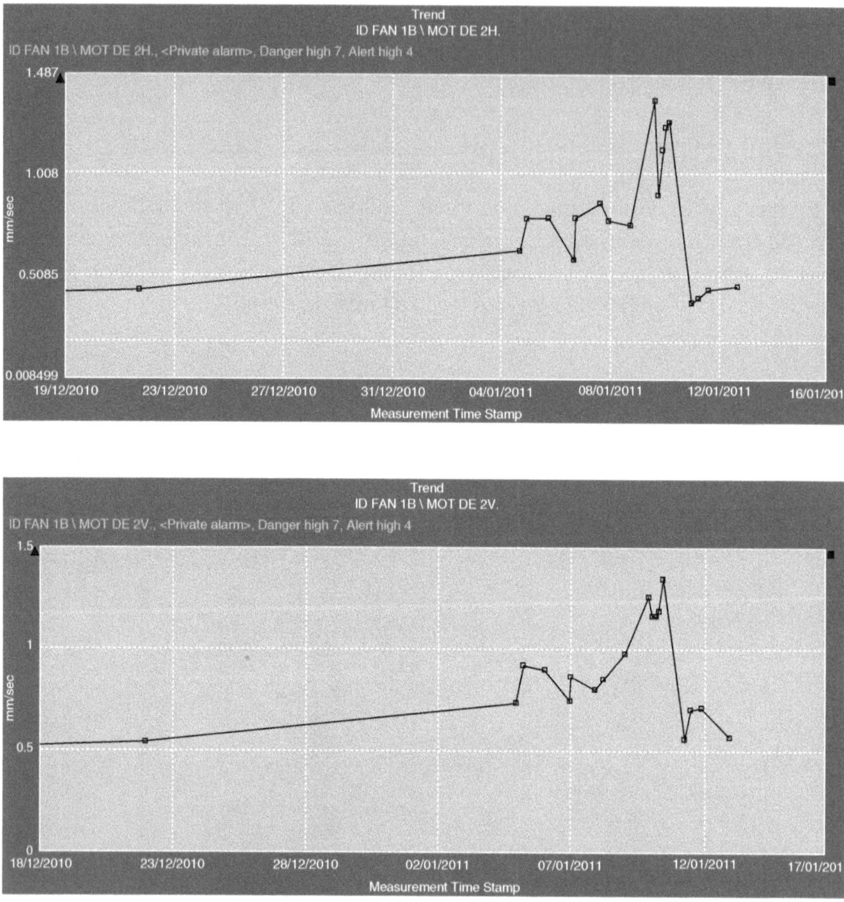

FIGURE 24.4 Vibration trend of ID Fan 1B motor DE (horizontal and vertical).

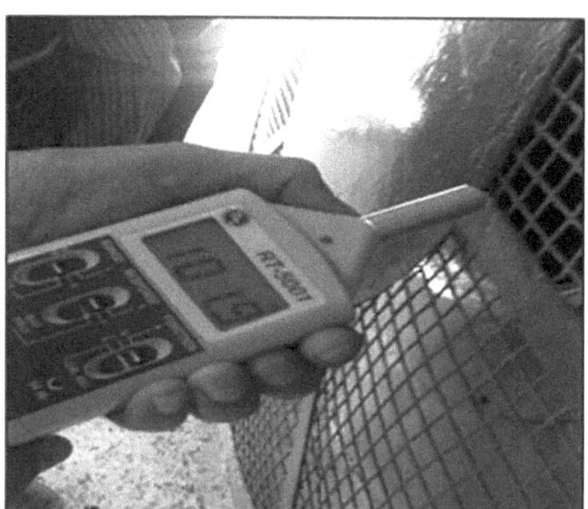

FIGURE 24.5 Noise collection at motor NDE side on 10 January 2011.

24.3 FREQUENCY ANALYSIS

Initially noise level at motor NDE side was increased from 94 to 101.9 dB. Normally, it remains in the range of 89–90 dB. Also, it was noticed that there was increase in the amplitude of vibration at motor NDE and DE bearing (horizontal and vertical) from 0.5 and 0.6 to 1.3 and 1.2 mm/s, respectively. Earlier it was found in the range of 0.5–0.6 mm/s. From Figures 24.6 to 24.8, spectrum analysis revealed that there was increase in multiples of running speed harmonics from 1×, 2×, 3× .... up to 10× frequency. It was suspected that motor bearings might loose on the shaft. Also, spectrums represent the bearing defect frequencies at Ball Spin Frequency (BSF), Ball Pass Frequency Outer race (BPFO) and Ball Pass Frequency Inner race (BPFI) (refer to Figure 24.9). These results clearly pinpoint the existence of bearing defects. Hence, it was concluded that vibrations of both motor bearings were increased due to bearing looseness on shaft and the deterioration of bearings was found in progress. So, it was decided to stop the fan for motor replacement.

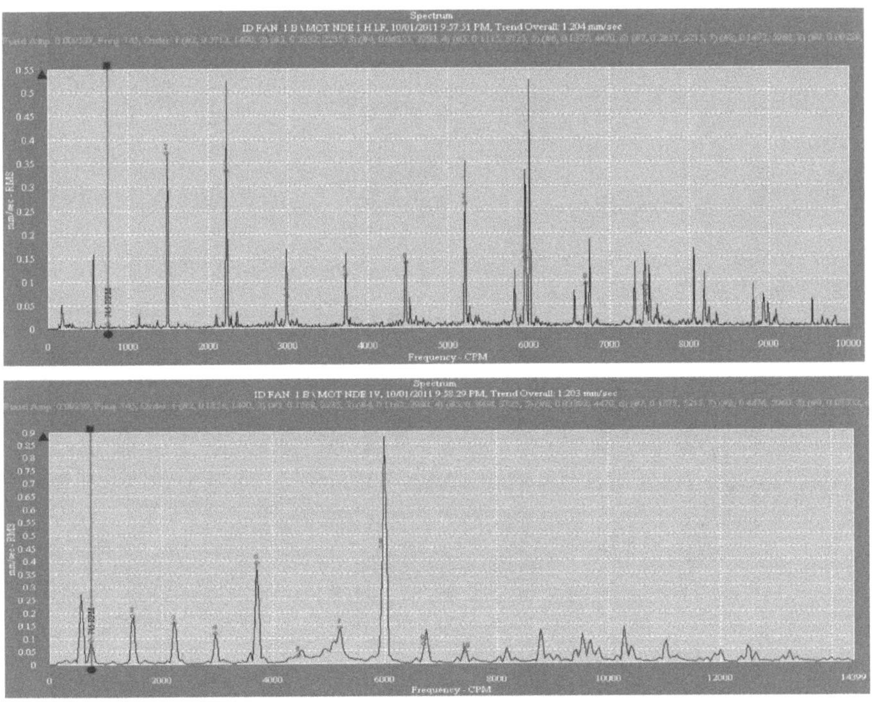

FIGURE 24.6 Vibration spectrum of motor NDE (horizontal and vertical) showing multiples of running speed frequency.

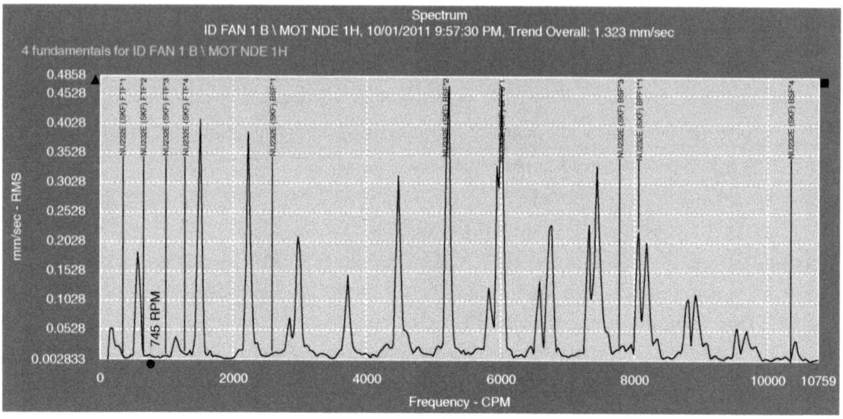

FIGURE 24.7 Vibration spectrum of motor NDE (horizontal) showing bearing defect frequencies.

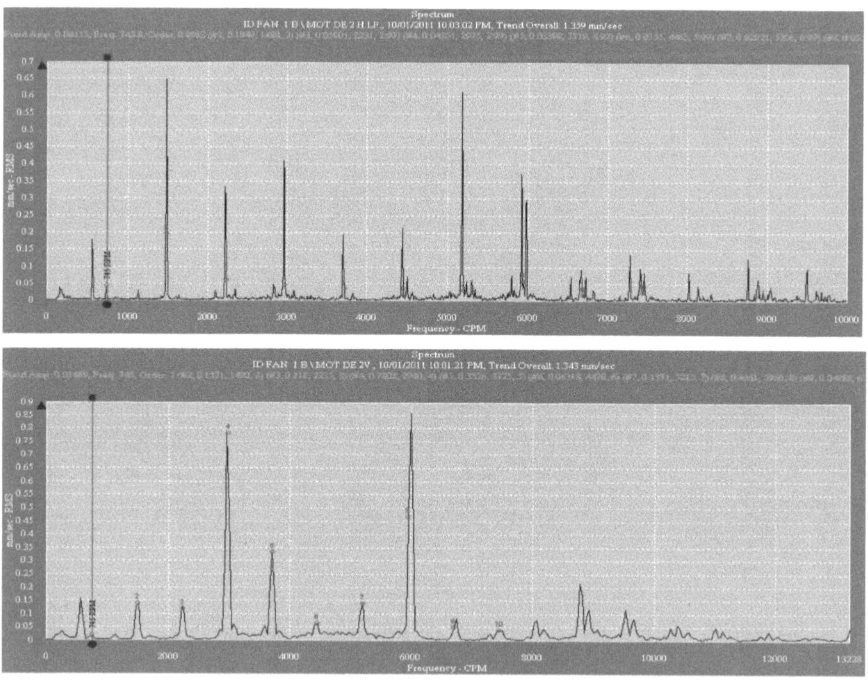

FIGURE 24.8 Vibration spectrum of motor DE (horizontal and vertical) showing multiples of running speed frequency.

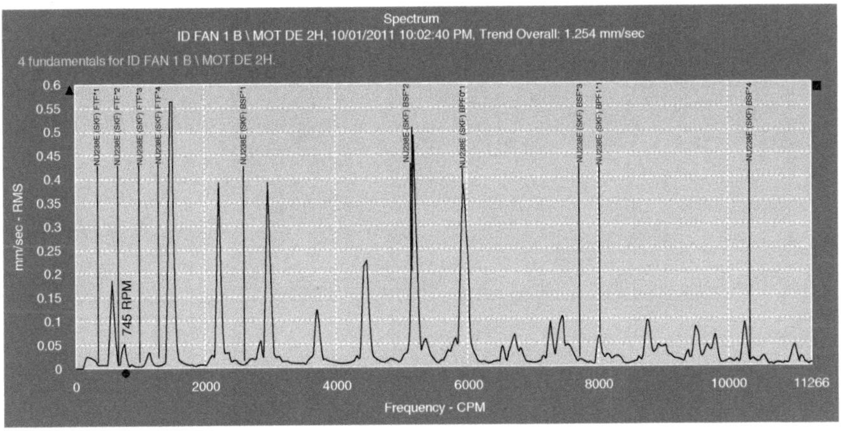

FIGURE 24.9 Vibration spectrum of motor DE (horizontal) showing bearing defect frequencies.

24.4 OBSERVATIONS RECORDED DURING MAINTENANCE WORK

Condition of cooling fan was found okay. The bearing clearances were found on higher side compared to original readings. During initial measurement, 0.5 mm shim of filler gauge was easily entering at NDE bearing inner race area which should not happen for healthy bearing (refer to Figure 24.10).

Condition of drive end (DE) bearing housing was found ok. Scoring marks were observed on non-drive end (NDE) side of bearing housing cap as shown in Figure 24.11.

Grease inside the NDE bearing was found blackish along with bearing burr particles, which were observed over stator rotor (see Figure 24.12). Inner race at NDE side bearing was cracked and got loosen over the shaft (see Figure 24.13). Scoring mark was observed over the shaft at NDE side due to relative motion between inner race of bearing and shaft. Due to this scoring mark, diameter of shaft was found as 159.34 mm at bearing area instead of 160.03 mm (refer to Figure 24.14). In addition, scoring mark was also observed on rollers of both DE and NDE bearings as shown in Figure 24.15. Detailed inspection of motor stator was carried out and found okay. No scoring marks were observed (refer to Figure 24.16). Detailed inspection of rotor was carried out and found okay. No scoring marks were observed on it (Figure 24.17).

FIGURE 24.10 Bearing clearance check.

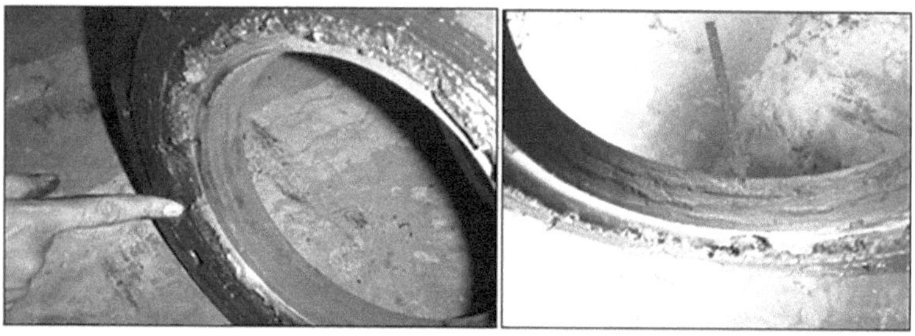

FIGURE 24.11 Scoring marks at non-drive end (NDE) side of bearing housing cap.

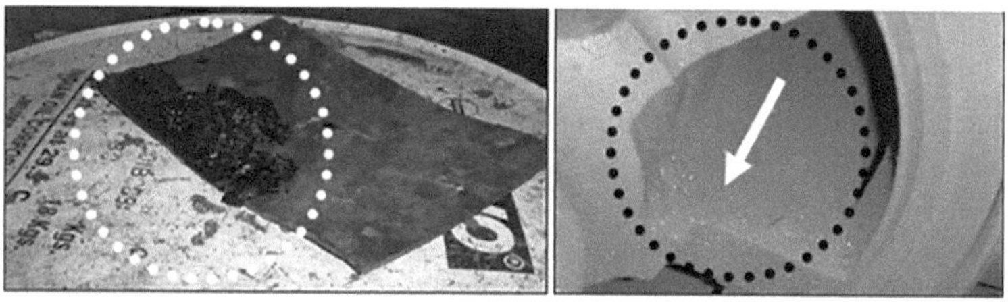

FIGURE 24.12 Bearing grease and wear particles.

FIGURE 24.13 Damaged NDE bearing inner race.

FIGURE 24.14 Shaft dimension comparison.

FIGURE 24.15 Scoring marks on both DE and NDE bearing rollers.

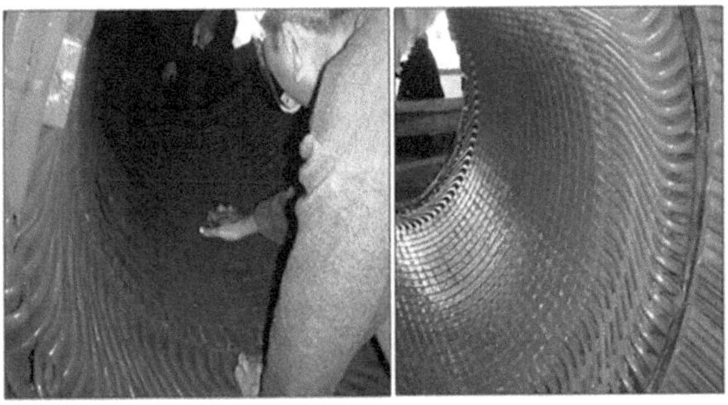

FIGURE 24.16 Inspection of motor stator.

FIGURE 24.17 Inspection of motor rotor.

24.5 MAINTENANCE WORK PERFORMED

Initially gear coupling from DE side was removed. Motor was replaced with the spare serviced motor. Then, gear coupling was mounted on motor shaft. Motor trial was taken at no load in decoupled condition, and vibrations were found in well within limit. No abnormal sound was noticed during the trial run of motor. ID fan alignment was done with rest of drive. ID fan boxed-up and made it available for operation. ID fan was taken in service. At full load, all vibrations and noise level were found in well within limit.

24.6 POST-MAINTENANCE OFFLINE VIBRATION READINGS

Table 24.1 and Figure 24.18 represent the comparison of vibration, acceleration and noise data for pre- and post-maintenance. All values were reduced to normal level after replacing the motor. It was concluded that the maintenance work was executed successfully.

TABLE 24.1
Vibration Data for Pre- and Post-Maintenance

EQUIPMENT : ID FAN 1B DATE : 10 to 11 January 2011

LOCATION: UNIT – 01

VIBRATION REPORT OF ID FAN 1B

MEASUREMENTS			OVERALL VIBRATION LEVELS		
Point		Position	Velocity (mm/s) RMS		
			Initial Readings	After Replacing Motor	
		DATE →	10/01/11	11/01/11	11/01/11
			Motor at Full Load	Decoupled Motor Trial	Motor at Full Load
Motor	(NDE)	H	**1.3 (1.0 g)**	0.3 (0.7 g)	**0.5 (0.6 g)**
		V	**1.2**	0.5	**0.6**
		A	**1.0**	0.3	**0.4**
Motor	(DE)	H	**1.2 (1.3 g)**	0.3 (1.2 g)	**0.4 (1.0 g)**
		V	**1.3**	0.4	**0.5**
		A	**1.0**	0.5	**0.4**
Hyd. Coup.	(DE)	H	1.3	–	1.2
		V	0.9	–	0.8
		A	2.2	–	2.0
Hyd. Coup.	(NDE)	H	2.2	–	1.9
		V	1.5	–	1.7
		A	2.1	–	1.7
Fan	(DE)	H	0.2	–	0.3
		V	0.2	–	0.3
		A	0.3	–	0.3
Fan	(NDE)	H	0.2	–	0.3
		V	0.3	–	0.3
		A	0.4	–	0.5

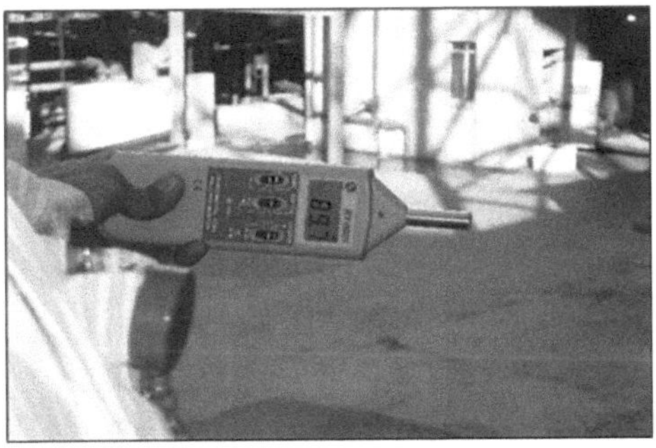

FIGURE 24.18 Post-maintenance noise level measurement.

24.7 POST-MAINTENANCE NOISE LEVEL AND VIBRATION SPECTRUMS

Noise level at motor NDE side was measured using Raython make handheld noise meter. Reduction in vibration level of motor NDE and DE bearing (horizontal and vertical) were observed from Figures 24.19 and 24.20.

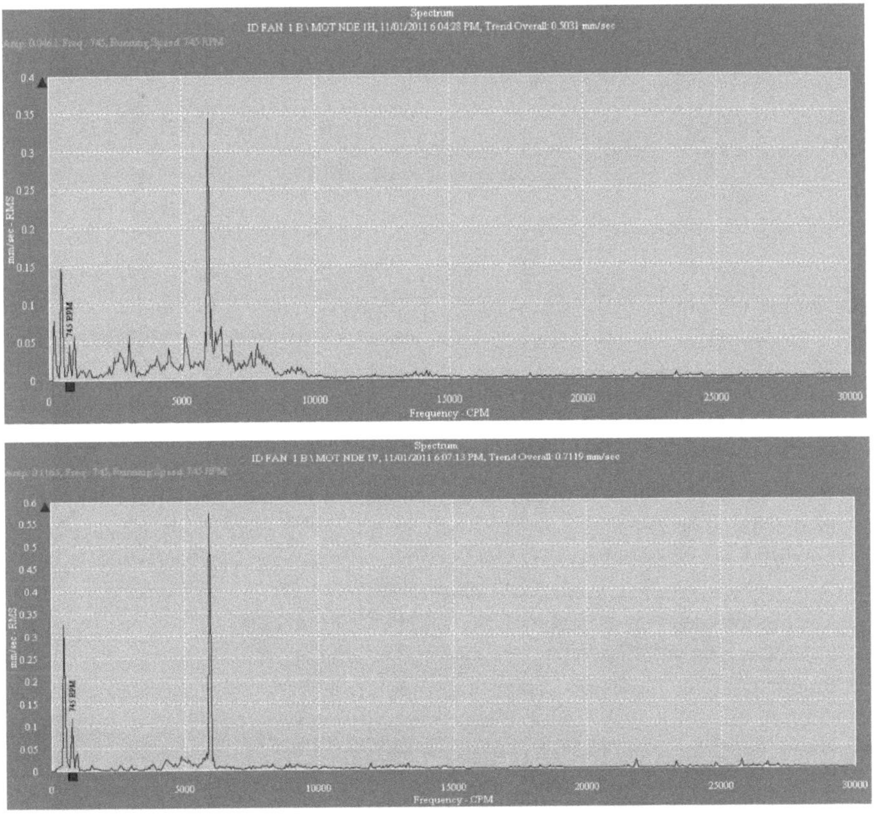

FIGURE 24.19 Post-maintenance vibration spectrum motor NDE (horizontal and vertical).

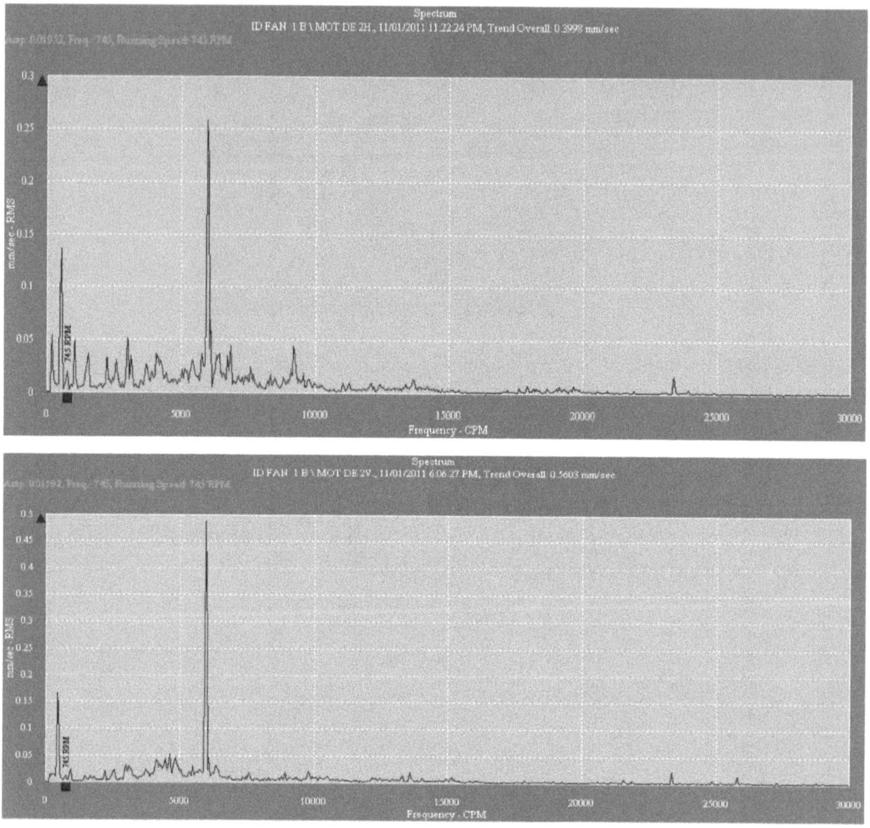

FIGURE 24.20 Post-maintenance vibration spectrum motor DE (horizontal and vertical).

Post-maintenance vibration spectrums (shown in Figures 24.19 and 24.20) were not showing any abnormality such as looseness and bearing defect frequencies after replacement of motor.

24.8 CONCLUSIONS

ID fan is critical equipment which is used to extract the flue gases out of furnaces through economiser, APH, stack and finally exit to atmosphere. It assists the FD fan by drawing out combustible gases from the furnace and maintains the partial negative pressure inside the boiler. Since ID fan has no standby, reliability of this fan needs to be very high. Boiler is provided with two numbers of induced draft fans. Each ID fan is provided with regulating damper control and scoop control for controlling the loading on fans. This made ID fan as a critical equipment in the system. If any fan motor fails during operation, it will lead to generation loss and in extreme case, tripping of plant and interruption in power supply to our valued customers of Mumbai, India.

This case study has clearly demonstrated that noise monitoring is valuable tool in rotating machinery analysis and is especially useful in detecting early and advanced bearing degradation even the vibration levels are measured on lower side. Each vibration analyst should use to improve vibration analysis accuracy because this can help maintenance organization to become more proactive and accurate.

The secondary damage to hydraulic coupling and fan bearing was avoided by using these techniques and other major benefit includes prevention of downtime, avoidance of major motor repair or replacement costs. The application of co-ordinated condition monitoring helped in diagnosing the exact root cause of abnormity well in advance and guided plant engineer to plan maintenance work during nearest available opportunity with no generation loss. This technique has saved the plant from huge losses in terms of generation, unplanned breakdown time and associated maintenance costs, which could have caused total cost of approx. 2.15 crores.

25 Case Study 24 *Boiler Feed Pump 1B: Misalignment of Booster Pump – January 2011*

EXECUTIVE SUMMARY

Following is a success story of the implementation of condition monitoring techniques at ADTPS, wherein imminent misalignment of HT auxiliary Boiler Feed Pump (BFP) – 1B due to damaged coupling bolts was diagnosed.

Vibration analysis helped in diagnosing the exact root cause of abnormity at early stage. This has saved the plant from huge losses in terms of generation, unplanned breakdown time and associated maintenance cost, which could have caused total cost of approximately Rs. 73 lakhs.

In December 2010, BFP service was done during unit overhaul as per the schedule. On 28 December 2010, during routine monitoring, it was observed that horizontal vibration of Booster Pump DE and NDE bearings were deviated from 2.9 to 4.8 and 2.8 to 4.1 mm/s (RMS), respectively even though the motor, hydraulic coupling and main pump vibration were found normal.

On 30 December 2010, horizontal vibration reading of Booster Pump DE was raised to 4.8–5.1. Then after, further raised to 5.9 mm/s on same day evening. In view of this increasing trend, it was decided to changeover the pump. While decoupling, it was found that six bolts of flexible coupling are sheared. In addition, foundation bolt shims below suction strainer were found loose. Rectification was carried out accordingly.

25.1 PRE-MAINTENANCE HISTORY

In December 2010, during unit overhaul, boiler feed pump (BFP) servicing was done as per the planned maintenance schedule. On 28 December 2010, during routine monitoring, it was observed that horizontal vibrations of booster pump drive end (DE) and non-drive end (NDE) bearing were deviated from 2.9 to 4.8 and 2.8 to 4.1 mm/s (RMS), respectively (Refer Figures 25.1 and 25.2) even though vibration readings of motor, hydraulic coupling and main pump was found normal.

Other parameters (i.e., pressure, temperature) of pump were found normal. It was decided to keep a close watch on vibration readings of booster pump. On 30 December 2010, horizontal vibration reading of booster pump DE was raised to 4.8 to 5.1 and up to 5.9 mm/s on same day. Due to such increasing trend of recorded vibrations, it was decided to change over the pump.

DOI: 10.1201/9781003432616-25

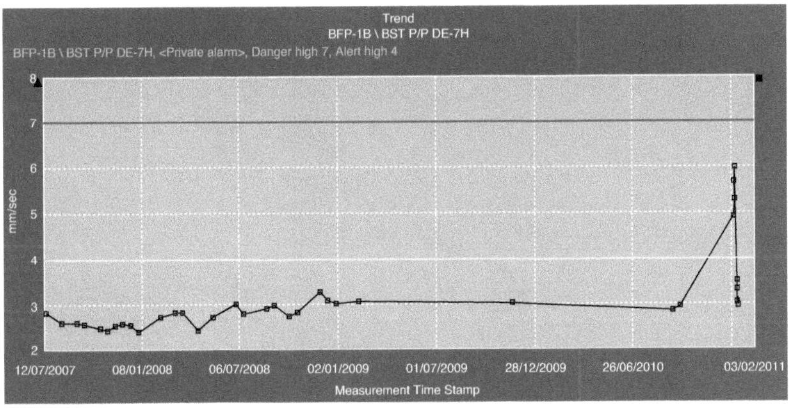

FIGURE 25.1 Vibration trend of booster pump DE bearing (horizontal).

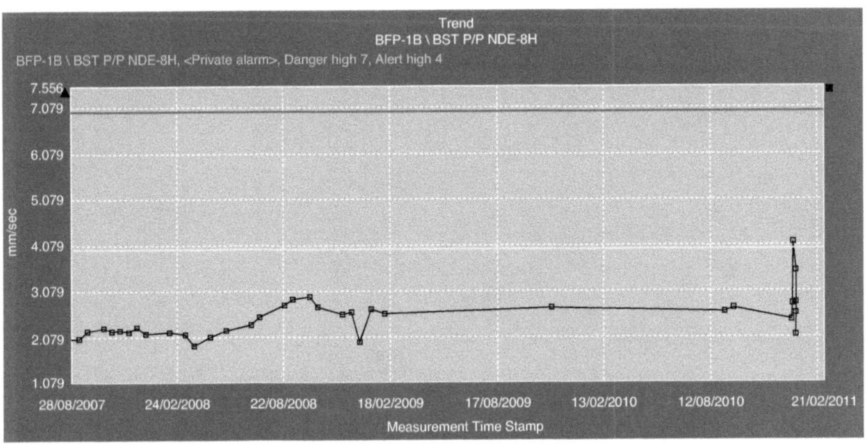

FIGURE 25.2 Vibration trend of booster pump NDE bearing (horizontal).

25.2 VIBRATION DIAGNOSIS

On 30 December 2010, horizontal vibrations of pump DE and NDE were reached to 5.9 and 4.1 mm/s, respectively (same readings remain normally in the range of 2.5–3.0 mm/s). The spectrum analysis for root cause was investigated as follows:

25.3 FREQUENCY ANALYSIS

Spectrum analysis (shown in Figures 25.3 and 25.4) revealed that there is increase in 1× frequency. It was suspected that booster pump was mis-aligned with motor. To confirm this misalignment, phase analysis was also carried out.

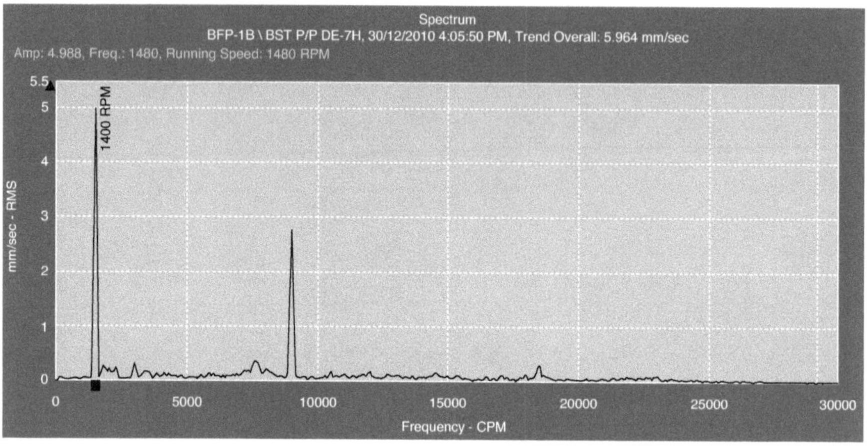

FIGURE 25.3 Vibration spectrum of booster pump DE (horizontal).

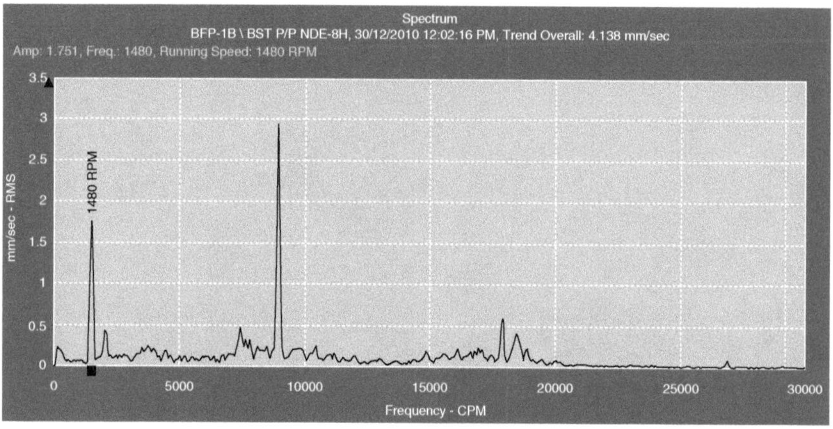

FIGURE 25.4 Vibration spectrum of booster pump NDE (horizontal).

Phase difference across coupling was measured 306° – 126° = 180° (refer to Figure 25.5). The above analysis shows that coupling phase difference was 180° (which pinpoints misalignment between booster pump and motor). Decision was taken to realign the booster pump with motor.

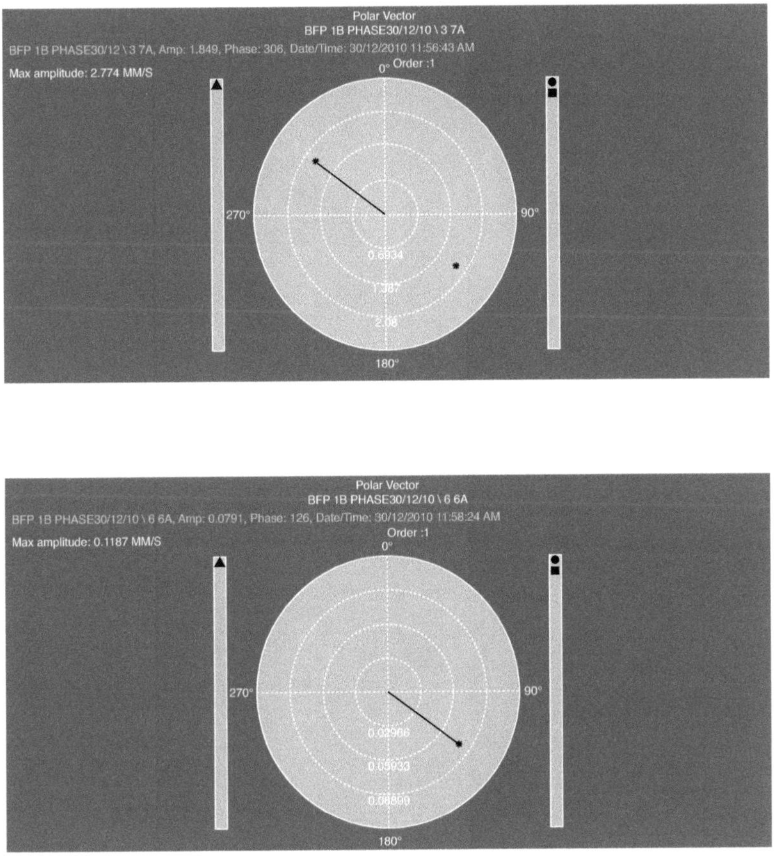

FIGURE 25.5 Phase analysis of booster pump DE (horizontal) and motor DE (horizontal).

25.4 OBSERVATIONS DURING MAINTENANCE WORK

No scoring marks were observed on both DE and NDE bearings. Condition of both bearing housings was found okay. While de-coupling the bolts of flexible coupling, six bolts were found sheared (see Figure 25.6). Shims (below suction strainer) were found loose as shown in Figure 25.7.

25.5 PERFORMED MAINTENANCE WORK

Booster pump was de-coupled from rest of the drive group. Before dismantling, initial alignment readings were taken for reference. New coupling bolts were replaced in place of sheared bolts. Five millimeter shims were added at all four corners of the suction strainer for smooth expansion. Alignment reading was taken and found to be disturbed.

Again 5 mm shims were removed, and 4 mm shims was added for precise alignment. Alignment was done with rest of group drive. Recorded readings were shown in Figure 25.8 (dial gauge on motor shaft and clamp on booster pump shaft). Trial of BFP was taken and all vibrations were found within limit. BFP boxed up and made it available for operation.

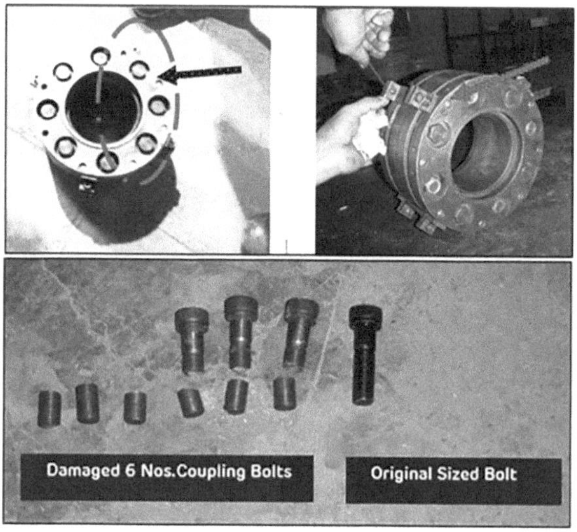

FIGURE 25.6 Damaged coupling bolts.

FIGURE 25.7 Loose shims at suction strainer.

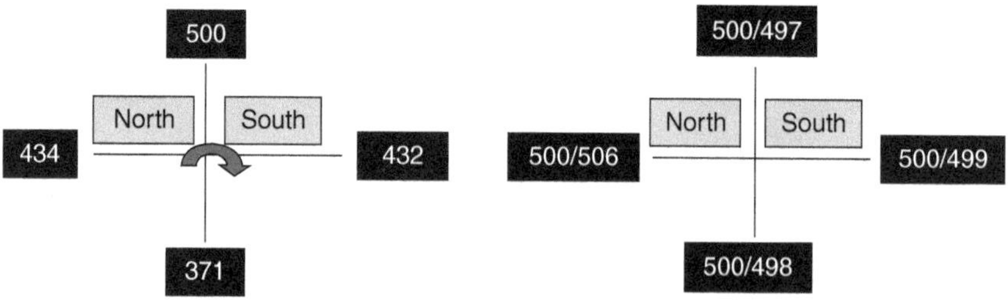

FIGURE 25.8 Initial and final alignment readings.

25.6 POST-MAINTENANCE OFFLINE VIBRATION READINGS

Table 25.1, Figures 25.9 and 25.10 represent comparison of vibration data for pre- and post-maintenance. All values reduced to normal level after replacing coupling bolts and alignment work. This was concluded that the maintenance work was executed successfully.

TABLE 25.1
Pre- and Post-Maintenance Vibration Data

EQUIPMENT : BOILER FEED PUMP – 1B DATE : 03.01.2011

LOCATION: TG BASEMENT UNIT 1

VIBRATION REPORT OF BOILER FEED PUMP – 1B

MEASUREMENTS			OVERALL VIBRATION LEVELS Velocity – mm/s (RMS)	
Point		Position DATE	INITIAL READINGS 30/12/2010	AFTER RE-ALIGNMENT 03/01/2011
Main		H	1.3	1.5
pump	**(NDE)**	V	1.6	1.7
		A	1.1	1.0
Main		H	2.1	1.9
pump	**(DE)**	V	2.4	2.4
		A	1.7	1.5
Hyd.		H	3.1	2.7
Coup.	**Pump**	V	3.3	3.4
		A	5.9	5.7
Hyd.		H	4.1	3.9
Coup.	**Motor end**	V	4.1	3.5
	end	A	5.8	5.9
Motor	**(Main pump end)**	H	1.0	1.1
	V	V	1.5	1.5
	A	A	0.4	0.4
Motor	**(Booster pump end)**	H	1.1	1.3
	V	V	1.2	1.3
	A	A	0.5	0.6
Booster	**(DE)**	H	**5.9**	**2.9**
pump	V	V	2.8	1.4
	A	A	2.0	1.3
Booster	**(NDE)**	H	**4.1**	**2.5**
pump	V	V	3.2	2.7
	A	A	1.2	1.1

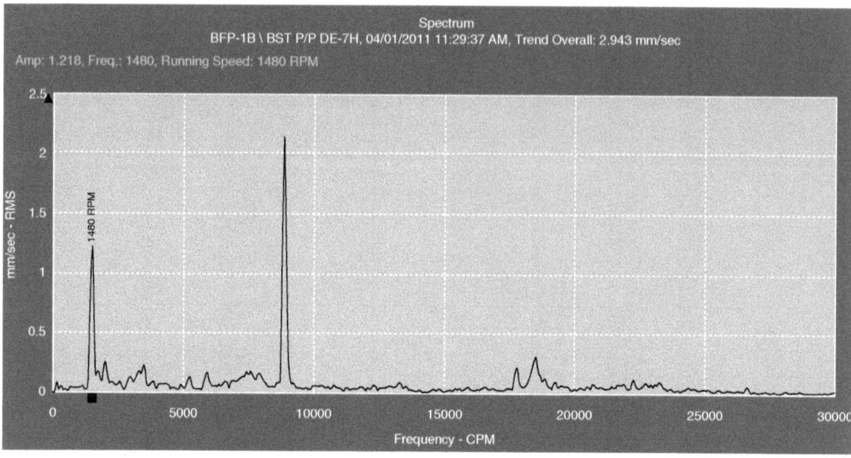

FIGURE 25.9 Post-maintenance vibration spectrum of booster pump DE bearing (horizontal).

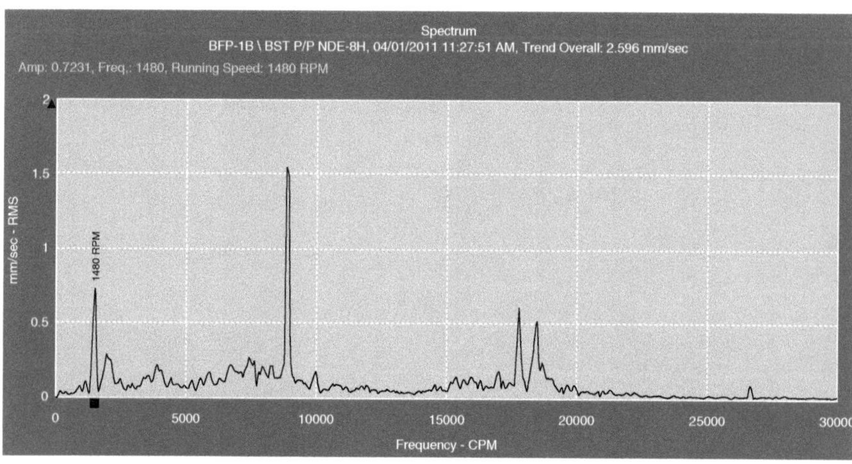

FIGURE 25.10 Post-maintenance vibration spectrum of booster pump NDE bearing (horizontal).

25.7 POST-MAINTENANCE VIBRATION SPECTRUMS

Above post-maintenance vibration spectrums were not showing any abnormality such as misalignment or looseness.

25.8 CONCLUSIONS

BFP is a critical auxiliary which is used to pump the feed water into the boiler. It is used to pump the feed water at high pressure to boiler drum placed at high altitude. The feed water pumping is done basically in two stages; first using a booster pump is provided as suction to the main pump. Since this pump supplies feed water to boiler, the reliability of pump needs to be very high. There are two numbers of BFP of 100% capacity (i.e., one is running at full load and other is in stand by condition). This provision is made as BFP which is critical equipment in the system and if it fails during non-availability of stand-by pump, it may lead to tripping of plant and interruption in power generation.

This case study clearly demonstrated that condition-based vibration testing is a vital component in rotating machinery analysis and detection of abnormality at early stage. Phase analysis played a major supportive role to identify the pinpoint misalignment.

The catastrophic failure of booster pump is avoided using these techniques and other major benefit includes prevention of downtime, avoidance of major pump repair or replacement costs. The importance of vibration and phase measurement helped in diagnosing the exact root cause of abnormity well in advance which could have caused total cost of Rs. 7,285,000.

26 Case Study 25
Coal Mill 1CD: Main Motor Bearing Damage – June 2008

EXECUTIVE SUMMARY

The present case study is a success story of the implementation of condition monitoring techniques at ADTPS, wherein imminent main motor DE bearing damage of HT auxiliary Coal Mill – 1CD was diagnosed.

This case study has clearly demonstrated that condition-based vibration testing is a vital component in rotating machinery analysis and detection of abnormality at early stage. Vibration monitoring and spectrum analysis helped in diagnosing the exact root cause of abnormity well in advance. This has saved the plant from huge losses in terms of generation, unplanned breakdown time and associated maintenance cost.

On 5 June 2008, during routine monitoring, it was observed that axial vibration of main motor NDE and DE bearings were deviated from 1.0 and 2.9 to 5.9 and 7.9 mm/s, respectively. No other deviation was noticed for main reducer gearbox and pinion NDE and DE bearing vibration. Spectrum analysis revealed predominant frequency of 7500 cpm, i.e., fifth running harmonic which suspected looseness of motor bearings. This spectrum compared with history data and found similar spectrum with high motor bearing axial vibration recorded on 27 February 1999 and at that time main motor was replaced.

By comparing history data and observing increasing trend in axial vibration of main motor NDE and DE bearings, it was decided to change over the coal mill for replacement of main motor DE bearing. During maintenance, it was observed that motor DE bearing clearances were on higher side and bearing found damaged. Detailed inspection of DE bearing casing, NDE bearing, coupling and motor shaft was carried out, and no scoring marks/abnormality was noticed.

26.1 PRE-MAINTENANCE HISTORY

On 5 June 2008, during routine monitoring, it was observed that axial vibrations of main motor NDE and DE bearing were deviated from 1.0 and 2.9 (RMS) to 5.9 and 7.9 mm/s (RMS), respectively (Refer Figures 26.1 and 26.2).

No vibration deviation was noticed for the main reducer gearbox and both pinion bearings. Also, coal mill-related process parameters (i.e. air flow, pressure and temperature) was found normal. Spectrum analysis revealed predominant frequency at 7500 cpm (i.e. fifth running harmonic) which suspected the looseness of motor DE bearing. This spectrum was compared with history data and found similar spectrum with high motor bearing axial vibration on 27 February 1999 and at that time main motor was replaced. Comparing historical data and by observing the increasing trend of axial vibration of main motor NDE and DE bearing, it was decided to change over the coal mill for replacement of main motor DE bearing.

 DOI: 10.1201/9781003432616-26

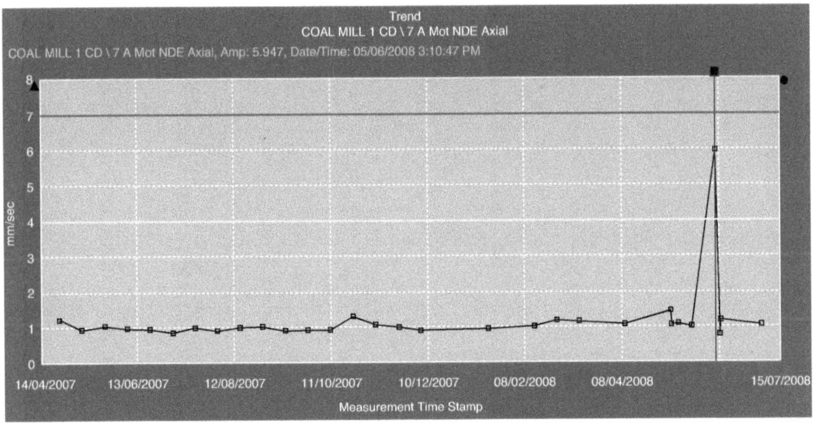

FIGURE 26.1 Vibration trend of main motor non drive end (NDE) bearing (axial).

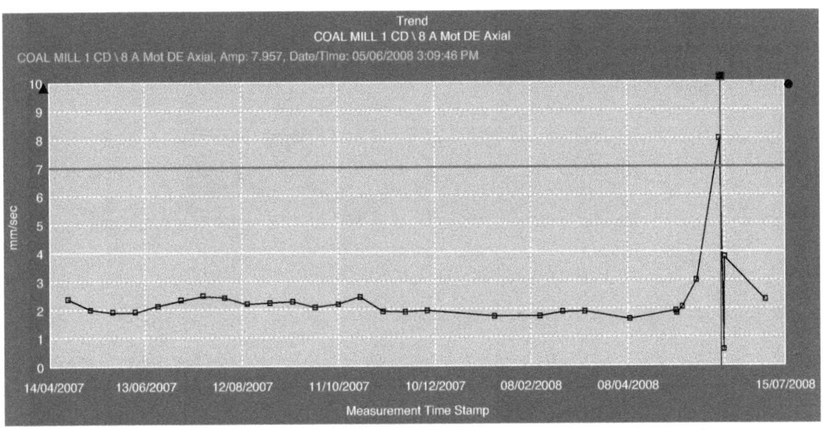

FIGURE 26.2 Vibration trend of main motor drive end (DE) bearing (axial).

26.2 VIBRATION DIAGNOSIS

The axial vibrations of both motor bearings were increased to 5.94 and 7.95 mm/s (NDE and DE). The spectrum analysis of root cause was investigated as follows:

The above both vibration spectrums (shown in Figures 26.3 and 26.4) show predominant frequency at 7500 CPM (i.e. fifth running harmonic) of motor speed which suspected looseness of motor bearings.

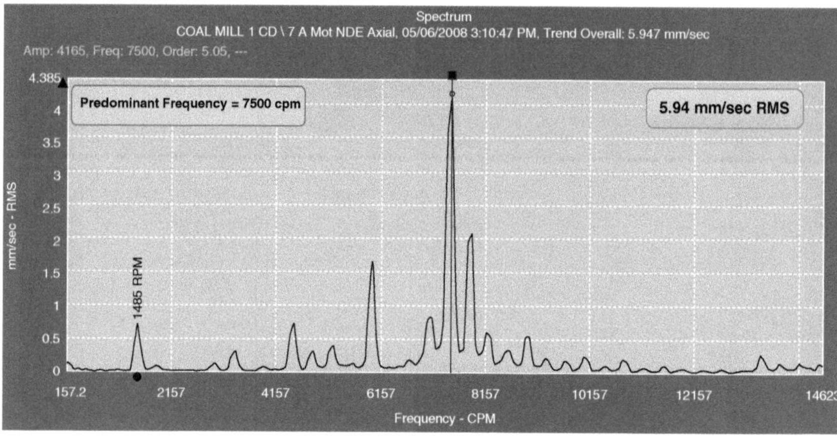

FIGURE 26.3 Vibration spectrum of main motor NDE bearing (axial).

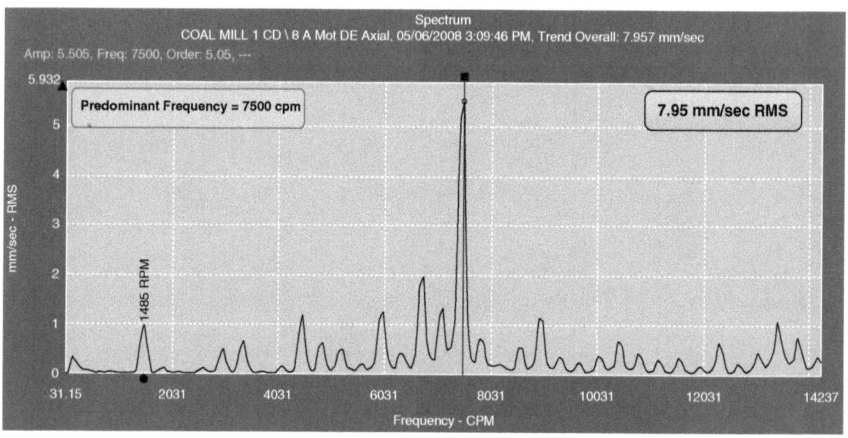

FIGURE 26.4 Vibration spectrum of main motor DE bearing (axial).

26.3 HISTORY DATA COMPARISON DATED 27 FEBRUARY 1999

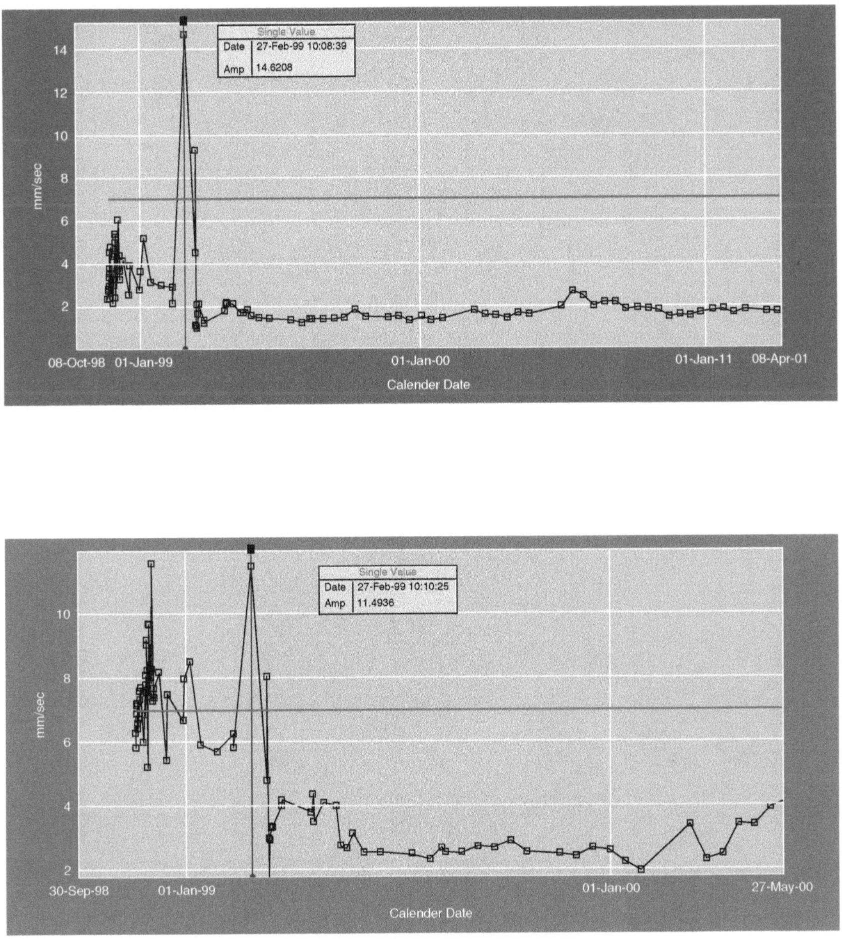

FIGURE 26.5 Vibration trend of main motor NDE and DE bearing (axial).

26.4 FREQUENCY ANALYSIS

From Figures 26.1 to 26.2, it was observed that there was an increase in vibration amplitude of main motor NDE and DE bearing only in axial direction from 1.0 and 2.9 to 5.9 and 7.9 mm/s (RMS), respectively. Earlier, it was found in the range of 1.0 and 3.0 mm/s. Spectrum analysis revealed predominant frequency at 7500 cpm (i.e. fifth running harmonic) of motor speed which suspected looseness of motor bearings. The vibration spectrum recorded on 5 June 2008 was compared with history data (Refer Figures 26.5–26.7) and found the same spectrum as captured on 27 February 1999. This confirmed that main motor bearing deterioration was in progress.

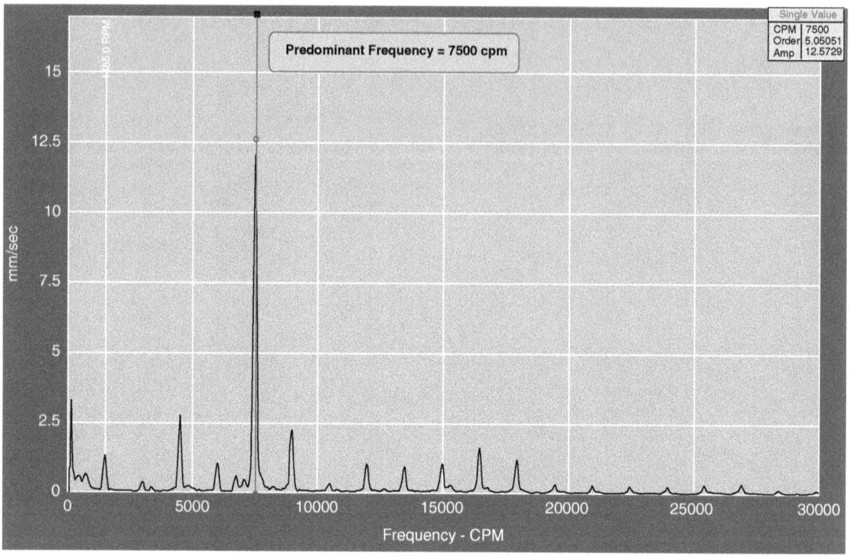

FIGURE 26.6 Vibration spectrum of main motor non drive end (NDE) bearing (axial).

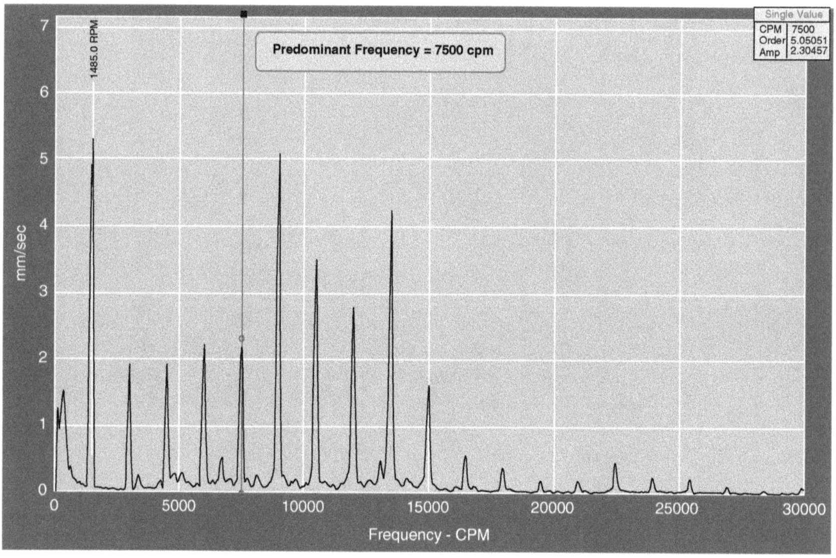

FIGURE 26.7 Vibration spectrum of main motor drive end (DE) bearing (axial).

26.5 OBSERVATIONS DURING MAINTENANCE WORK

Motor DE bearing clearances was checked and found on higher side (as shown in Figure 26.8). Damaged main motor DE bearing (top and bottom half) was shown in Figure 26.9. Motor NDE bearing clearances was checked and found within limit (see Figure 26.10). No scoring marks were observed on NDE side bearing (see Figure 26.11). No damage/abnormality was noticed to bearing casing and motor shaft (Figure 26.12).

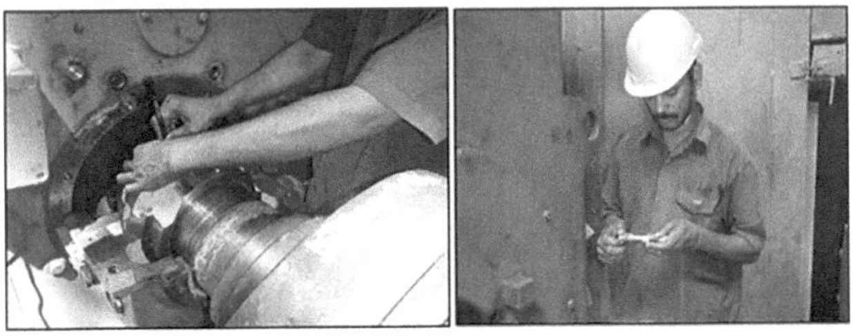

FIGURE 26.8 Main motor DE bearing clearance check.

FIGURE 26.9 Damaged motor DE bearing.

FIGURE 26.10 Proactive check for main motor NDE bearing.

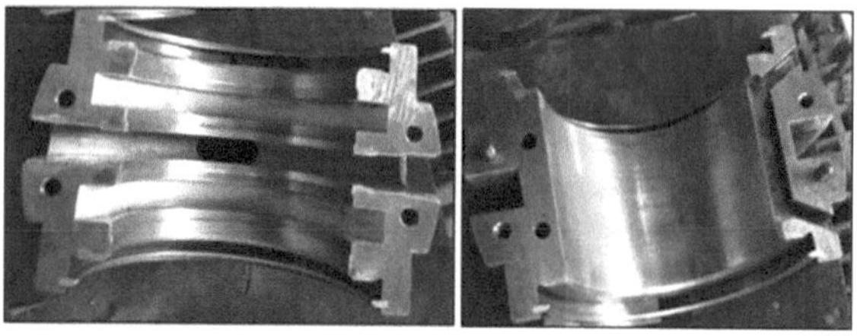

FIGURE 26.11 Motor NDE bearing top (LHS) and bottom (RHS) half.

FIGURE 26.12 Motor NDE bearing casing and shaft.

26.6 PERFORMED MAINTENANCE WORK

Coal mill main motor was de-coupled from rest of the group drive (see Figure 26.13). Before dismantling, initial alignment readings were taken for reference (see Figure 26.14). Main motor DE bearing was removed as shown in Figure 26.15. After blue matching, a new bearing of main motor DE was installed as shown in Figure 26.16. Main motor was re-aligned with rest of drive (see Figure 26.17). Coal mill trial was taken at full load and all vibrations readings were found within limit. Coal mill boxed up and made it available for operation.

FIGURE 26.13 Main motor de-coupled from rest of drive.

FIGURE 26.14 Pre-maintenance alignment data collection.

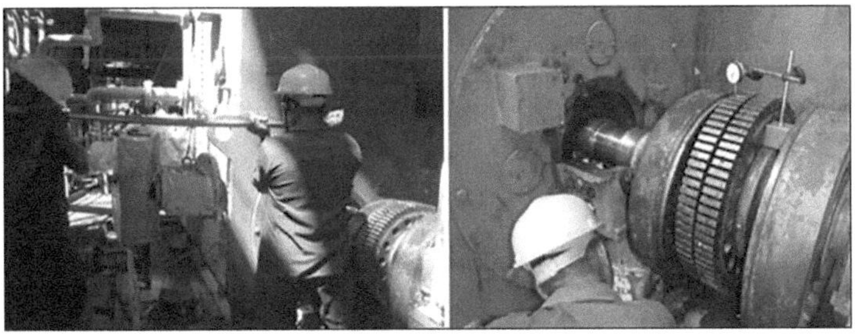

FIGURE 26.15 Removal of damaged motor DE bearing.

FIGURE 26.16 Installation of new motor DE bearing.

FIGURE 26.17 Final alignment work.

26.7 POST-MAINTENANCE OFFLINE VIBRATION READINGS

Table 26.1, Figures 26.18 and 26.19 represent the comparison of vibration data for pre- and post-maintenance. All values were reduced to normal level after replacing main motor DE bearing and alignment work.

TABLE 26.1
Pre-and Post-Maintenance Vibration Data

EQUIPMENT : COAL MILL 1 CD DATE : 5 to 8 JUNE 2008

LOCATION: - UNIT – 1

MEASUREMENTS			OVERALL VIBRATION LEVELS			
Point		Position	Velocity (mm/s) RMS			
			Initial Readings		After Replacing Motor DE Bearing	
		DATE →	5 June 2008	6 June 2008	7 June 2008	8 June 2008
			Motor at Full Load	Decoupled Motor	Decoupled Motor	Motor at Full Load
Main motor	(NDE)	H	1.51	0.30	0.25	0.86
		V				
		A				
		V	1.85	0.46	0.46	1.44
		A	**5.94**	1.19	1.00	**1.19**
Main motor	(DE)	H	3.28	0.35	0.27	2.10

(Continued)

TABLE 26.1 (*Continued*)
Pre-and Post-Maintenance Vibration Data

EQUIPMENT : COAL MILL 1 CD DATE : 5 to 8 JUNE 2008

LOCATION: - UNIT – 1

MEASUREMENTS			OVERALL VIBRATION LEVELS			
Point		Position	Velocity (mm/s) RMS			
			Initial Readings		After Replacing Motor DE Bearing	
		DATE →	5 June 2008	6 June 2008	7 June 2008	8 June 2008
			Motor at Full Load	Decoupled Motor	Decoupled Motor	Motor at Full Load
		V	3.67	0.59	0.51	2.84
		A	**7.95**	0.55	0.55	**3.81**
	(DE)	H	5.67	-	-	3.68
GB		V	3.74	-	-	3.97
		A	4.50	-	-	3.58
	(NDE)	H	5.50	-	-	3.98
GB		V	5.14	--	-	5.21
		A	5.08	-	-	3.99
Bearing		H	4.78	-	-	3.97
		V	3.16	-	-	3.12
		A	5.01	-	-	3.91
Bearing		H	4.92	-	-	5.31
		V	5.92	-	-	6.27
		A	5.50	-	-	3.61
Pinion	(DE)	H	2.88	-	-	2.55
		V	5.39	-	-	5.22
		A	3.05	-	-	1.29
Pinion	(NDE)	H	1.70	-	-	2.33
		V	4.62	-	-	4.93
		A	2.10	-	-	1.91

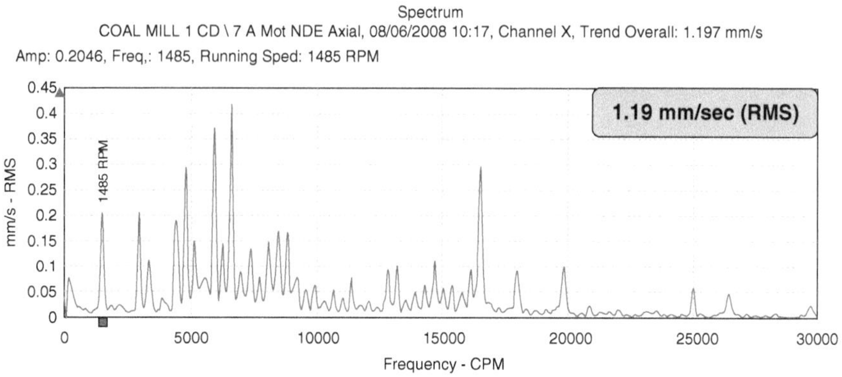

FIGURE 26.18 Vibration spectrum for post-maintenance main motor NDE bearing (axial).

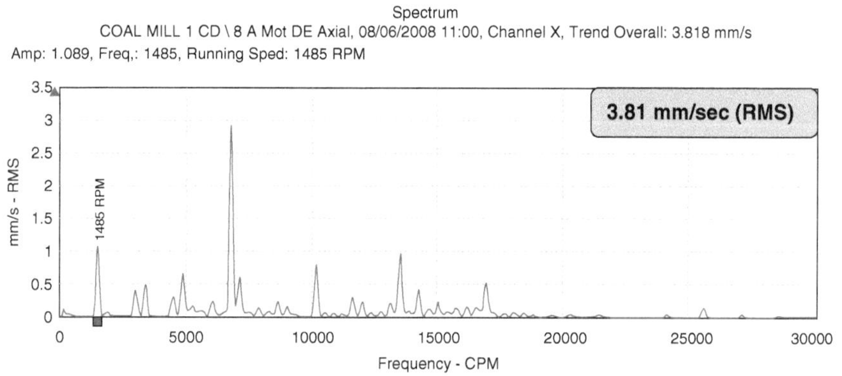

FIGURE 26.19 Vibration spectrum for post-maintenance main motor DE bearing (axial).

26.8 POST-MAINTENANCE VIBRATION SPECTRUMS

Above post-maintenance vibration spectrum did not show predominant frequency at 7500 cpm and any abnormality such as bearing damage or looseness.

26.9 CONCLUSIONS

Coal mill is an important auxiliary, which is used to crush the coal in required size and feed in the boiler. Since mill supplies coal as fuel to boiler, the reliability needs to be very high. It is a critical auxiliary in the process, and its failure directly affects the generation loss of 0.125 MUs/h which costs revenue of around Rs. 437,500 per hour. There are three coal mills of 50% capacity (i.e. two running at full load and other is in stand by condition). This provision is made because coal mill is critical equipment in the system and if it fails during non-availability of stand-by mill, it will lead to tripping of plant and interruption in power supply to our valued customers of Mumbai.

This case study has clearly demonstrated that condition-based vibration testing is a vital component in rotating machinery analysis and detection of abnormality at early stage. Vibration spectrum analysis helped in diagnosing the exact root cause of abnormity well in advance and its history helped us to detect the bearing failure and hence the unavailability situation of coal mill to operation is avoided. Each vibration analyst should analyze history data to improve analysis accuracy that can help maintenance organization to become more proactive and accurate. The catastrophic failure of motor was avoided using these techniques, and other major benefits include prevention of downtime, avoidance of major motor repair or replacement costs. The importance of vibration and history data helped in diagnosing the exact root cause of abnormity well in advance which could have caused total cost of Rs. 151,300.

Index

Note: **Bold** page numbers refer to tables and *italic* page numbers refer to figures.

FSC
www.fsc.org
MIX
Papier | Fördert gute Waldnutzung
FSC® C083411

Kolibri 360 GmbH
Ferdinand-Jühlke-Straße 7
99095 Erfurt, Deutschland
produktsicherheit@kolibri360.de